S0-ATZ-781

This book
belongs to

....

S0-ATZ-781

AMAZING
Animal
facts

Amazing
Animal
facts

Written by
Rupert Matthews, and Gerald Legg

Edited by
Fiona Corbridge, Philippa Moyle, Hazel Songhurst, and Nicola Wright

zigzag

CONTENTS

FANTASTIC SEA CREATURES 6

Coral creatures • Seashore creatures • Deep-sea creatures
Cool creatures • Microscopic creatures • Giant creatures
Coastline creatures • Clever creatures • Deadly creatures
Flying creatures • Scaly creatures • Strange creatures
Prehistoric creatures • Mysterious creatures

CREEPY CRITTERS 34

Growers • Developers • Hunters and trappers
Rotters, tunnellers and burrowers • Suckers • Flyers
Crawlers and runners • Hoppers, jumpers, and skaters
Slitherers and wrigglers • Camourflagers • Tricksters
Flashers and warners • Singers and glowers • Carers

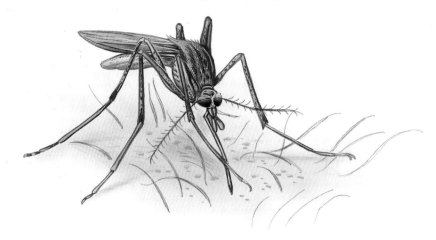

MONSTER ANIMALS 62

DANDEROUS AND DEADLY 90

Coral creatures

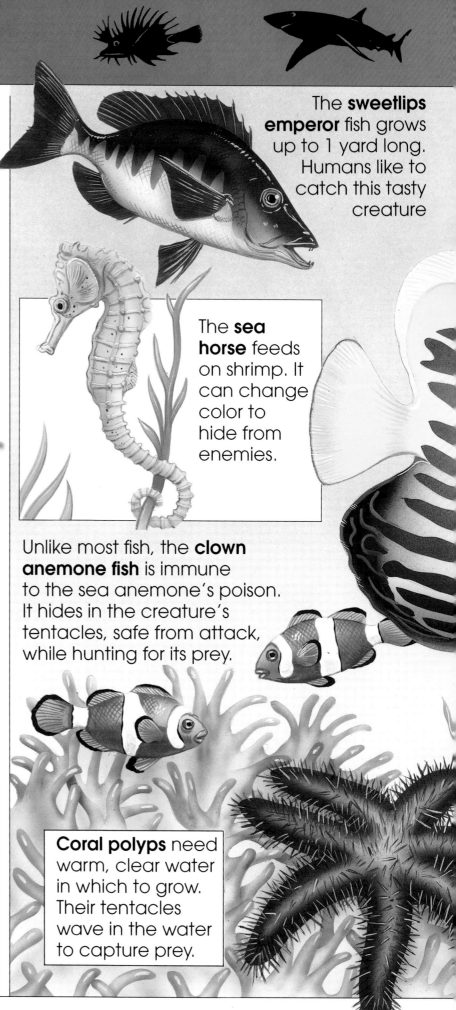

Coral reefs are made from millions of tiny creatures called coral polyps. When it dies, each polyp leaves behind a tiny limestone skeleton. There are thousands of types of polyp.

Many other creatures live on the reef. The shallow water and rocky crevices provide an ideal home.

The largest coral reef in the world is the **Great Barrier Reef** off the east coast of Queensland, Australia. It is over 1,243 mi. long.

The **sweetlips emperor** fish grows up to 1 yard long. Humans like to catch this tasty creature

The **sea horse** feeds on shrimp. It can change color to hide from enemies.

Unlike most fish, the **clown anemone fish** is immune to the sea anemone's poison. It hides in the creature's tentacles, safe from attack, while hunting for its prey.

Coral polyps need warm, clear water in which to grow. Their tentacles wave in the water to capture prey.

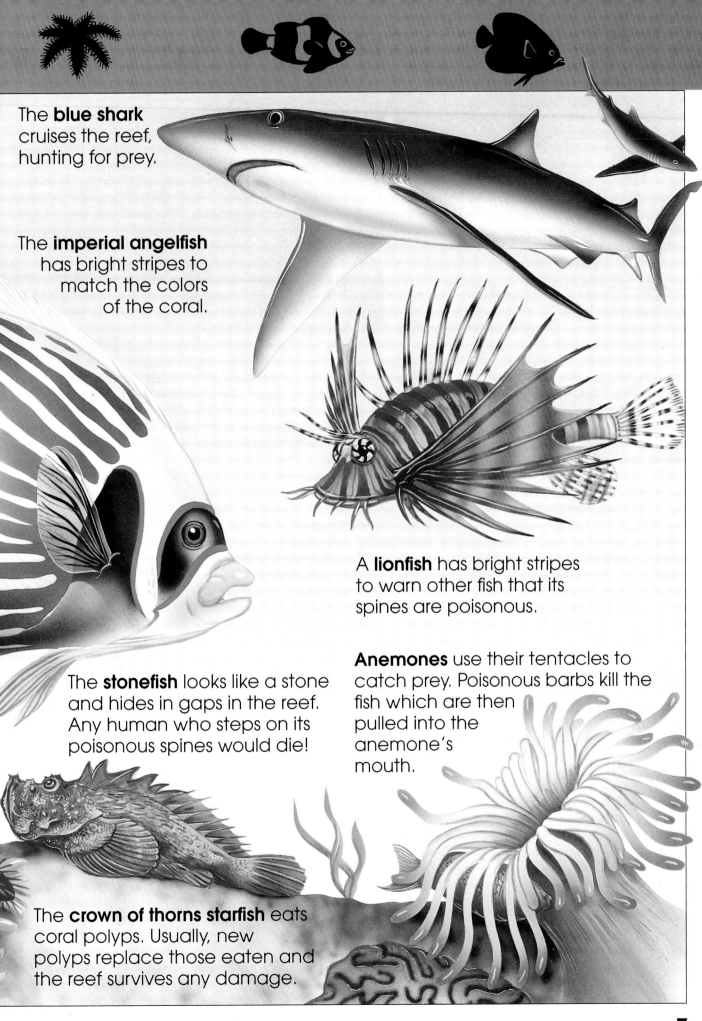

The **blue shark** cruises the reef, hunting for prey.

The **imperial angelfish** has bright stripes to match the colors of the coral.

A **lionfish** has bright stripes to warn other fish that its spines are poisonous.

Anemones use their tentacles to catch prey. Poisonous barbs kill the fish which are then pulled into the anemone's mouth.

The **stonefish** looks like a stone and hides in gaps in the reef. Any human who steps on its poisonous spines would die!

The **crown of thorns starfish** eats coral polyps. Usually, new polyps replace those eaten and the reef survives any damage.

Life on the shore can be very difficult for animals. As the tide comes in and goes out, their surroundings change from dry land to shallow sea.

Shrimp feed among the seaweed. When the tide goes out, they swim into deeper water, but sometimes they are caught in rock pools.

Pounding waves throw animals around. The sand is always moving as the sea pushes and pulls it around. Seashore animals must be tough to survive.

The **scorpion fish** and other kinds of small fish feed among the stones in rock pools. They swim out with the tide.

The **masked crab** lives on sandy beaches. When the tide goes out, it burrows into the sand. The tips of its two antennae poke out of the sand and act as breathing tubes.

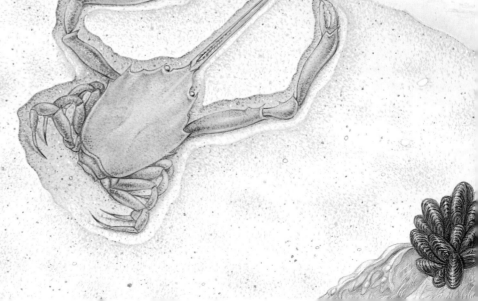

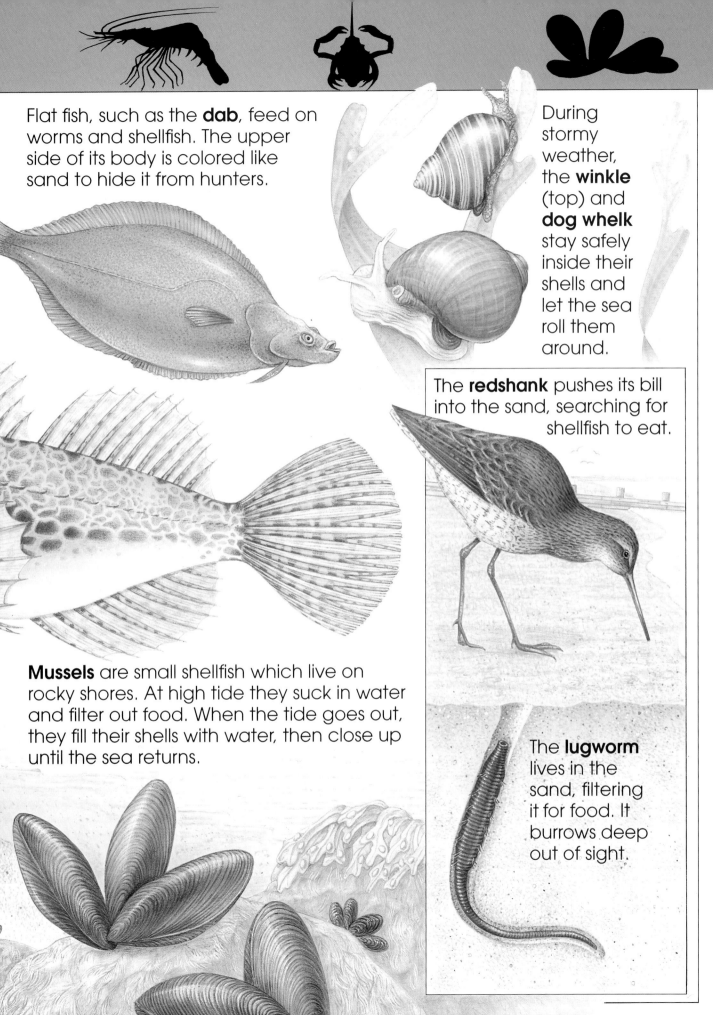

Flat fish, such as the **dab**, feed on worms and shellfish. The upper side of its body is colored like sand to hide it from hunters.

During stormy weather, the **winkle** (top) and **dog whelk** stay safely inside their shells and let the sea roll them around.

The **redshank** pushes its bill into the sand, searching for shellfish to eat.

Mussels are small shellfish which live on rocky shores. At high tide they suck in water and filter out food. When the tide goes out, they fill their shells with water, then close up until the sea returns.

The **lugworm** lives in the sand, filtering it for food. It burrows deep out of sight.

9

Deep-sea creatures

Most sea creatures live near the surface, where the water is warm and sunlit. The light cannot travel very deep and the sea's currents rarely move the warm surface water down to the depths of the ocean.

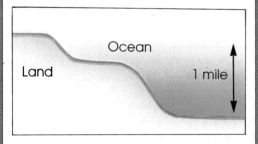

Ocean

Land

1 mile

At depths of more than 1 mile, the sea is very cold and completely dark. Some very strange creatures live here. They feed on each other and on food which drifts down from above.

Sperm whales dive down for food.

The **giant squid** can grow to 65 feet long.

Deep-sea shrimp can glow to attract a mate.

The **deep-sea angler fish** has a long growth over its mouth which glows faintly. This attracts other fish, which are then swallowed whole!

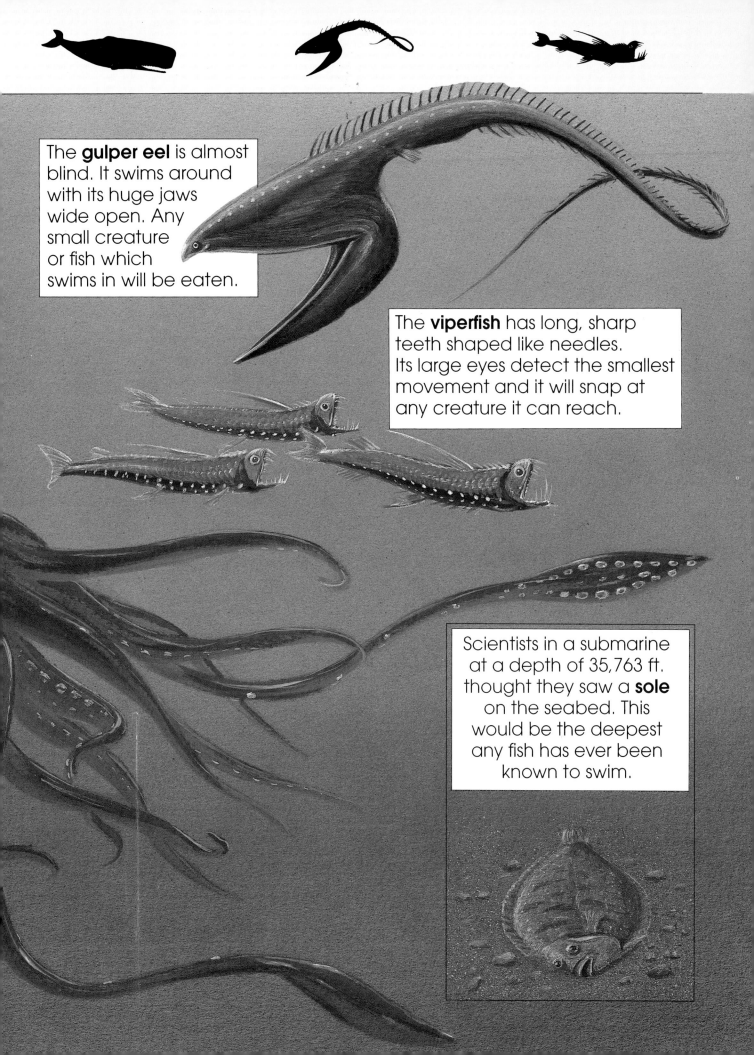

The **gulper eel** is almost blind. It swims around with its huge jaws wide open. Any small creature or fish which swims in will be eaten.

The **viperfish** has long, sharp teeth shaped like needles. Its large eyes detect the smallest movement and it will snap at any creature it can reach.

Scientists in a submarine at a depth of 35,763 ft. thought they saw a **sole** on the seabed. This would be the deepest any fish has ever been known to swim.

Around both the North and South Poles, the weather is very cold. A layer of ice floats on top of the sea all year round.

Animals which live there must be able to keep warm. They may have thick fur, like the polar bear, or layers of fat under their skin, like the common seal.

Killer whales prey on any creatures they can catch. They will even push ice from underwater to knock penguins and seals into the sea.

Penguins are birds that live around the South Pole. They lay their eggs on the ice and hunt for fish in the sea.

The largest penguin is the **Emperor penguin** which grows to over 3 feet tall.

The smallest penguin is the **fairy penguin**, which is only 16 in. tall.

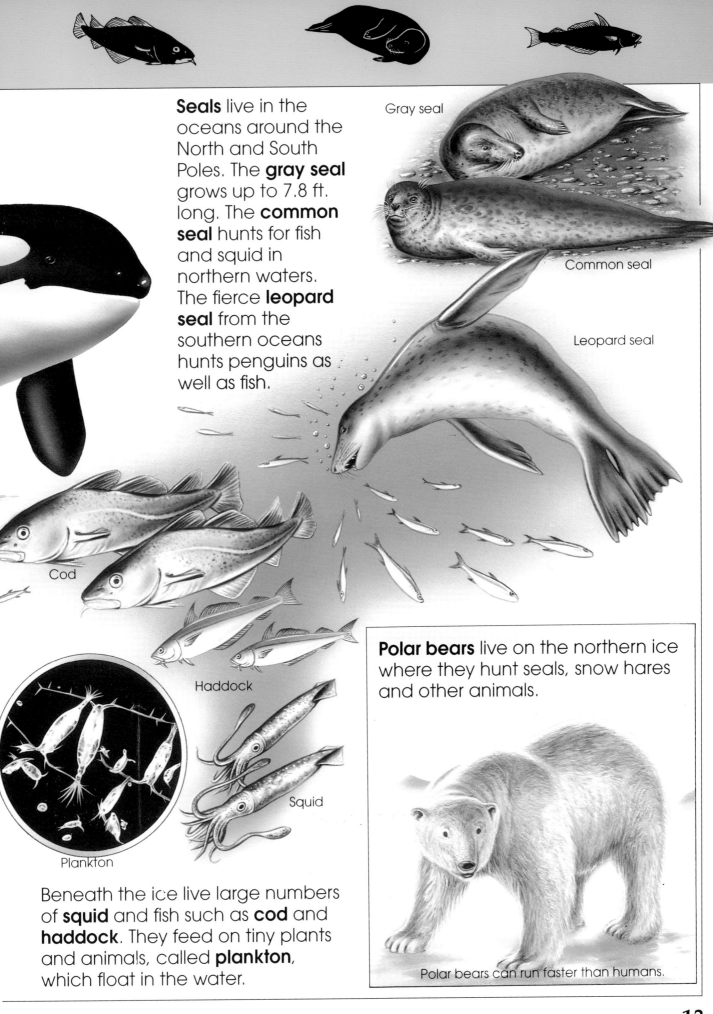

Seals live in the oceans around the North and South Poles. The **gray seal** grows up to 7.8 ft. long. The **common seal** hunts for fish and squid in northern waters. The fierce **leopard seal** from the southern oceans hunts penguins as well as fish.

Gray seal

Common seal

Leopard seal

Cod

Haddock

Squid

Plankton

Beneath the ice live large numbers of **squid** and fish such as **cod** and **haddock**. They feed on tiny plants and animals, called **plankton**, which float in the water.

Polar bears live on the northern ice where they hunt seals, snow hares and other animals.

Polar bears can run faster than humans.

13

The smallest living things in the sea are called plankton. They are so small that you could fit 40,000 of them on the end of your thumb.

Plankton can be either plants or animals. They are food for the larger sea animals.

Large clouds of plankton drift in the surface waters of all seas.

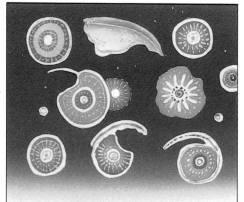

Phytoplankton are microscopic plants. They use the sunlight's energy to grow like plants on land.

The smallest animals are made of a single bodycell. **Ceratium** moves by thrashing a long, whip-like "arm." It feeds on tiny plants.

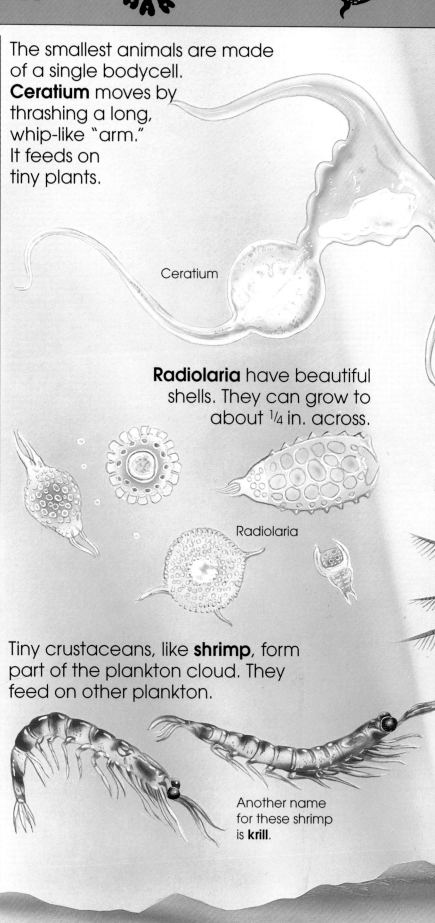

Ceratium

Radiolaria have beautiful shells. They can grow to about ¼ in. across.

Radiolaria

Tiny crustaceans, like **shrimp**, form part of the plankton cloud. They feed on other plankton.

Another name for these shrimp is **krill**.

Some of the **plankton** are the young of much larger creatures. Because they drift with the ocean currents, these creatures can travel much further than they can as adults, allowing them to reach new homes.

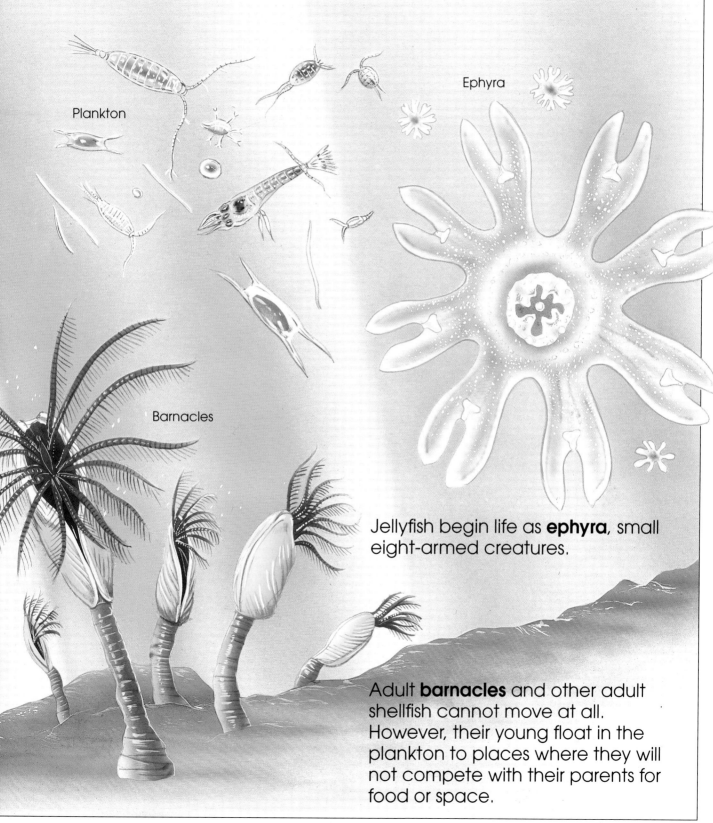

Plankton

Ephyra

Barnacles

Jellyfish begin life as **ephyra**, small eight-armed creatures.

Adult **barnacles** and other adult shellfish cannot move at all. However, their young float in the plankton to places where they will not compete with their parents for food or space.

Whales are mammals which have evolved to live in the sea. They have fins instead of legs and a powerful tail to push them through the water.

The **bowhead whale's** head is 20 ft. long. This is one-third of its length. The jaws are packed with baleen to filter food from the seawater.

The largest whales feed on plankton. They have special filters, called baleen, in their mouths which strain seawater and remove the tiny animals and plants to be eaten.

Like all mammals, whales breathe air, so they need to come to the surface from time to time.

The earliest known whale is **basiliosaurus**, which lived about 40 million years ago.

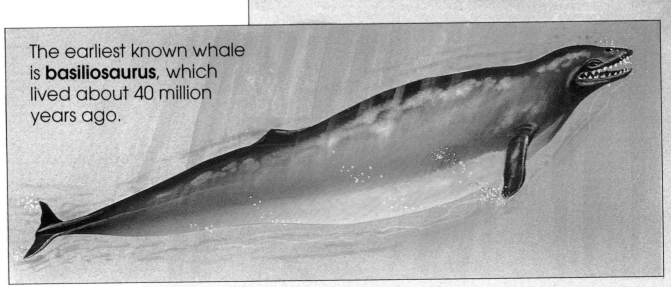

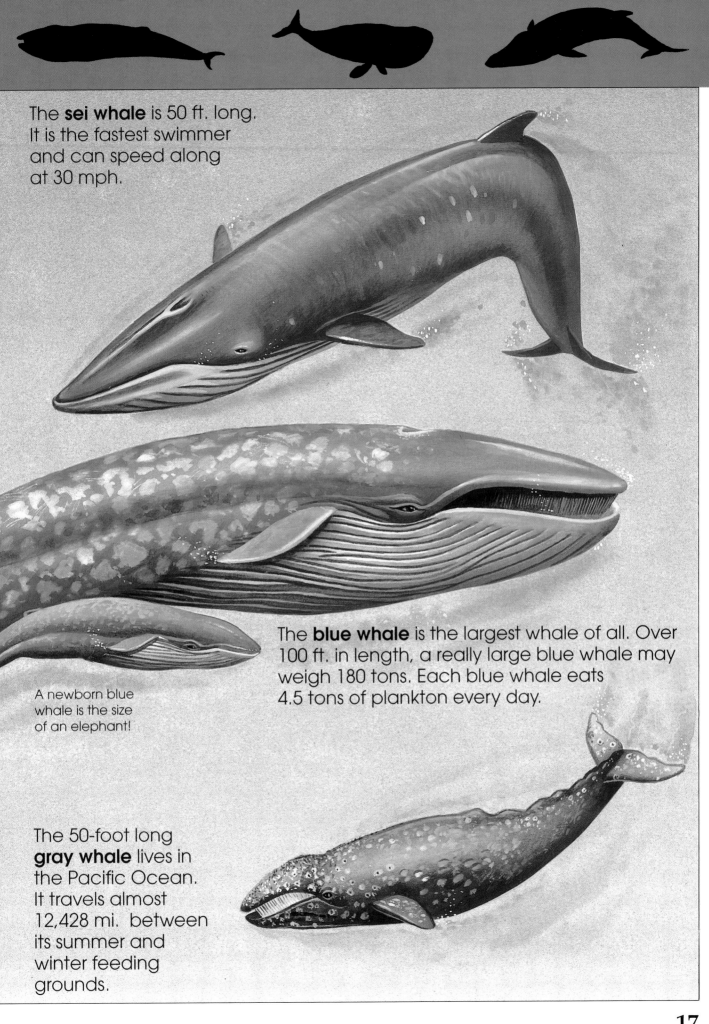

The **sei whale** is 50 ft. long. It is the fastest swimmer and can speed along at 30 mph.

A newborn blue whale is the size of an elephant!

The **blue whale** is the largest whale of all. Over 100 ft. in length, a really large blue whale may weigh 180 tons. Each blue whale eats 4.5 tons of plankton every day.

The 50-foot long **gray whale** lives in the Pacific Ocean. It travels almost 12,428 mi. between its summer and winter feeding grounds.

17

Seals are mammals which have evolved to live in the oceans.

Their legs have become flippers to help them swim, but they can still move on land.

Seals spend some of their time on shore, either caring for their babies or resting from hunting for fish.

Elephant seals were once hunted for the rich oil their bodies contain. At one time, only about a hundred were still alive, but today there are over 50,000 of them.

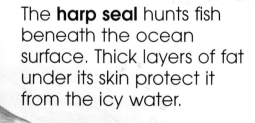

The **harp seal** hunts fish beneath the ocean surface. Thick layers of fat under its skin protect it from the icy water.

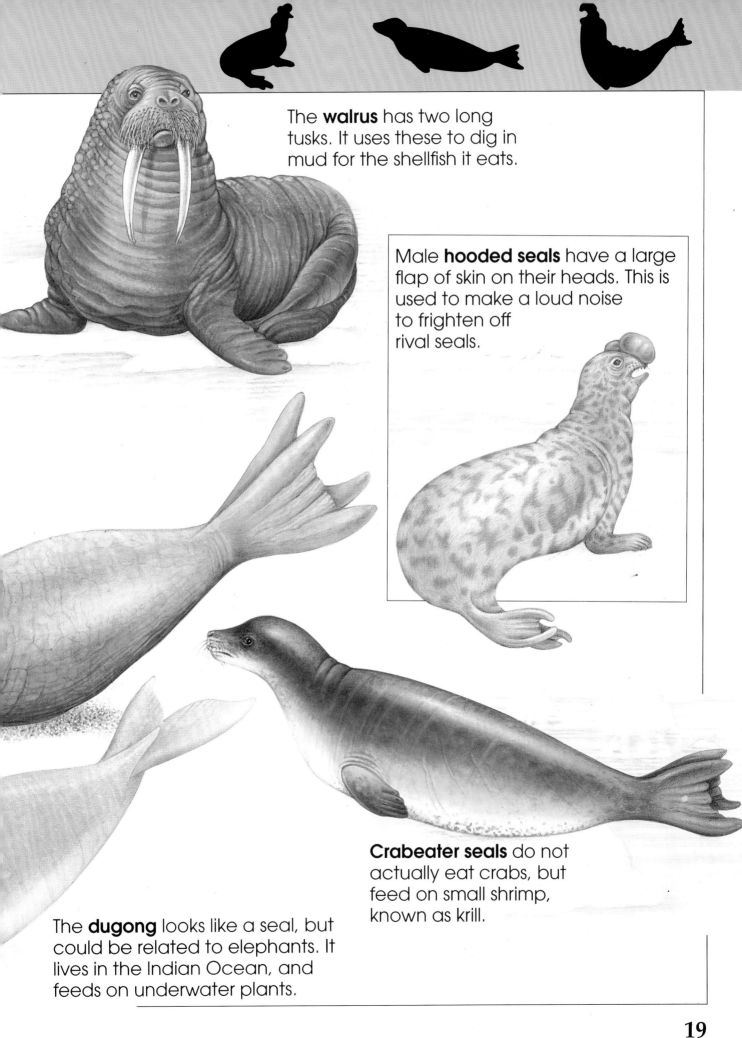

The **walrus** has two long tusks. It uses these to dig in mud for the shellfish it eats.

Male **hooded seals** have a large flap of skin on their heads. This is used to make a loud noise to frighten off rival seals.

Crabeater seals do not actually eat crabs, but feed on small shrimp, known as krill.

The **dugong** looks like a seal, but could be related to elephants. It lives in the Indian Ocean, and feeds on underwater plants.

19

Dolphins belong to a group of whales called toothed whales. They do not eat plankton but hunt squid and fish.

Dolphins are very intelligent creatures. They communicate with each other using different sounds arranged like words in a sentence.

Some dolphins are very rare. The **shepherd's beaked whale** is a recent discovery.

Dolphins are social animals. They live in family groups. If one dolphin is sick or injured, others will come to its rescue.

Dolphins are mammals. They breathe air and feed their babies with milk.

The **bouto** is a dolphin that lives in the Amazon River.

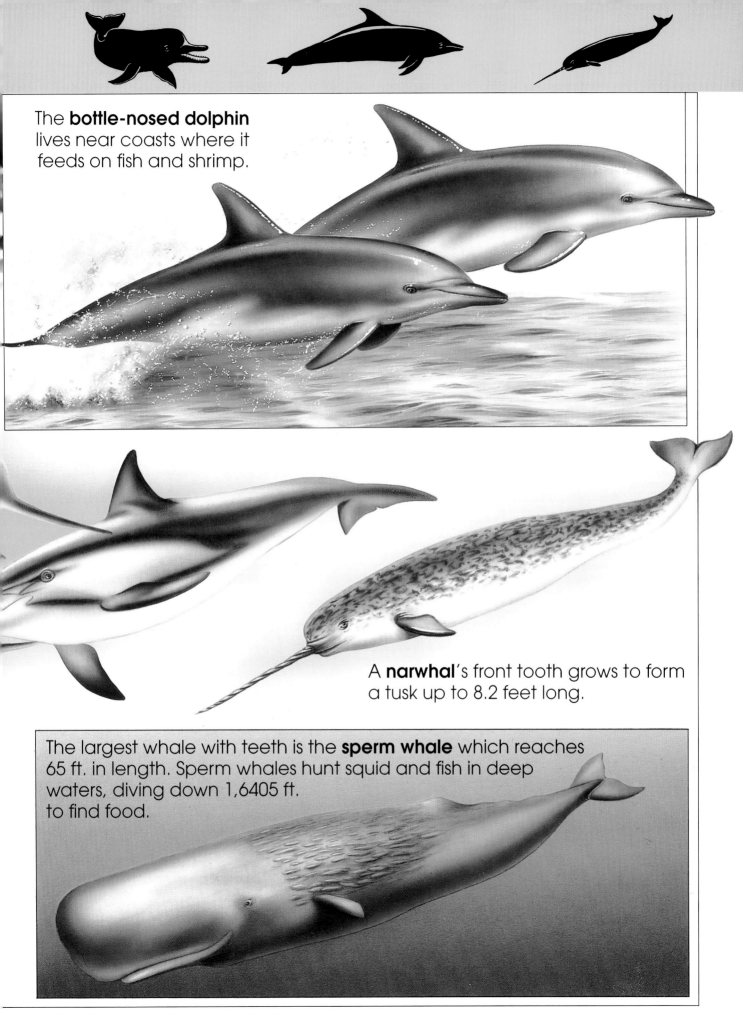

The **bottle-nosed dolphin** lives near coasts where it feeds on fish and shrimp.

A **narwhal**'s front tooth grows to form a tusk up to 8.2 feet long.

The largest whale with teeth is the **sperm whale** which reaches 65 ft. in length. Sperm whales hunt squid and fish in deep waters, diving down 1,6405 ft. to find food.

Deadly creatures

Sharks and rays are found in all seas. Their skeletons are made of soft cartilage instead of hard bone.

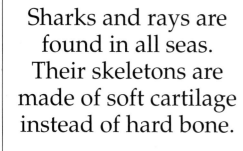

Sharks hunt other sea creatures, using their sharp teeth and strong muscles to overpower their prey.

The largest shark is the 60-foot long **whale shark**. Unlike most sharks, it does not hunt other animals. Instead, it feeds on plankton.

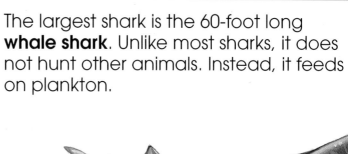

The harmless **whale shark** lives in warm waters.

The **great white shark** sometimes attacks people.

The largest hunting shark is the **great white shark**. It may grow to 23 ft. long and usually feeds on larger fish and other animals.

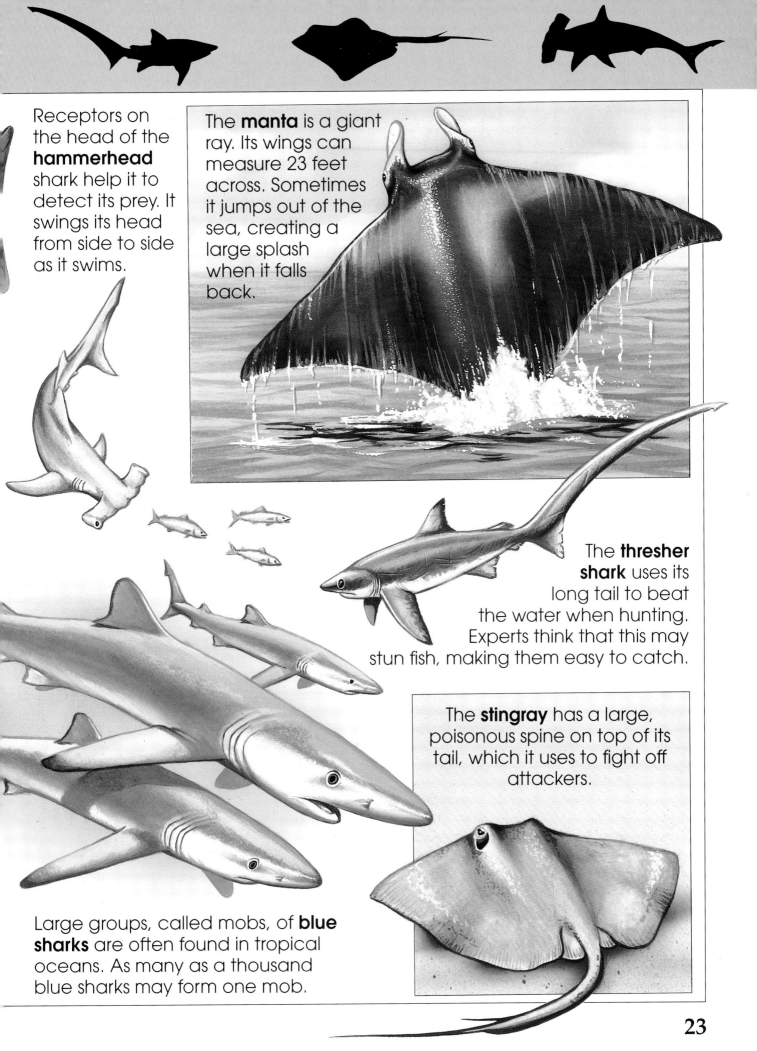

Receptors on the head of the **hammerhead** shark help it to detect its prey. It swings its head from side to side as it swims.

The **manta** is a giant ray. Its wings can measure 23 feet across. Sometimes it jumps out of the sea, creating a large splash when it falls back.

The **thresher shark** uses its long tail to beat the water when hunting. Experts think that this may stun fish, making them easy to catch.

The **stingray** has a large, poisonous spine on top of its tail, which it uses to fight off attackers.

Large groups, called mobs, of **blue sharks** are often found in tropical oceans. As many as a thousand blue sharks may form one mob.

Flying creatures

Many birds live at sea feeding on fish or other sea creatures.

Most sea birds nest on islands, where their eggs and young are safe from attack.

Sea birds often make long journeys between their nesting sites and feeding grounds. Arctic terns travel between the Arctic and the Antarctic.

The **great skua** is a large bird, over 20 in. long. It hunts other sea birds, as well as fish.

Herring gulls are very common. They feed on fish and shrimp, but will also fly inland to raid garbage dumps and picnic areas.

A **skimmer** finds fish by flying just above the surface of the sea, with its bill in the water. As soon as the bill strikes a fish, it is snapped up.

The largest sea bird is the **wandering albatross**, which has wings 11 feet across. Long ago, sailors believed it was bad luck to kill an albatross.

Steamer ducks live around the coast. They cannot fly, but swim along the shore looking for shellfish, shrimp and crabs to eat.

Puffins nest on cliffs and rocky islands. The females lay just one egg each year.

Gannet fly around searching for fish in the water. They may dive from a height of 100 ft. to catch their prey.

Reptiles are animals such as lizards. Most live on land.

A few types of reptile have evolved to live in the ocean, but they need to come to the surface often to breathe air.

Most sea reptiles lay their eggs on dry land. They may come ashore once a year to do this.

The **green turtle** has a tough shell to protect it from attack. It feeds on seaweed and jellyfish.

The **leatherback turtle** has no shell, but the skin on its back is very thick and tough.

Estuary crocodiles live off the coasts of northern Australia. They can grow to be over 20 ft. long and are the largest sea reptiles alive today.

Ridley turtles crunch up shellfish with their strong jaws.

The **banded sea snake** lives in the Pacific Ocean where it hunts fish. It is one of the most poisonous snakes in the world.

The **hawksbill turtle** is very rare. Not long ago, it was hunted for its shell. This was used to make things such as ornate boxes and spectacle frames.

Marine iguanas live around the remote Galapagos Islands in the Pacific Ocean. They dive into the ocean to feed on seaweed.

Marine iguanas come ashore to bask in the sun.

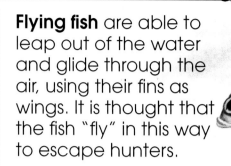

There are many fish in the oceans that look strange to us, but they are actually very well adapted to their surroundings.

Thousands of fish have evolved to live in different places - on coral reefs, in icy waters, near the surface of the sea, or on the seabed

Flying fish are able to leap out of the water and glide through the air, using their fins as wings. It is thought that the fish "fly" in this way to escape hunters.

When danger threatens, the **porcupine fish** gulps huge amounts of water and swells up to four times its usual size. The stiff spines stick out to make the fish look like a spiky football.

The **four-eyed fish** swims at the surface with each of its two eyes half in and half out of the water. The fish looks for insect prey on the surface, while watching for danger under the sea.

The **swordfish** has a bony upper jaw which can be over 1 yard long and shaped like a sword. Nobody knows what the sword is used for.

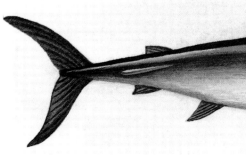

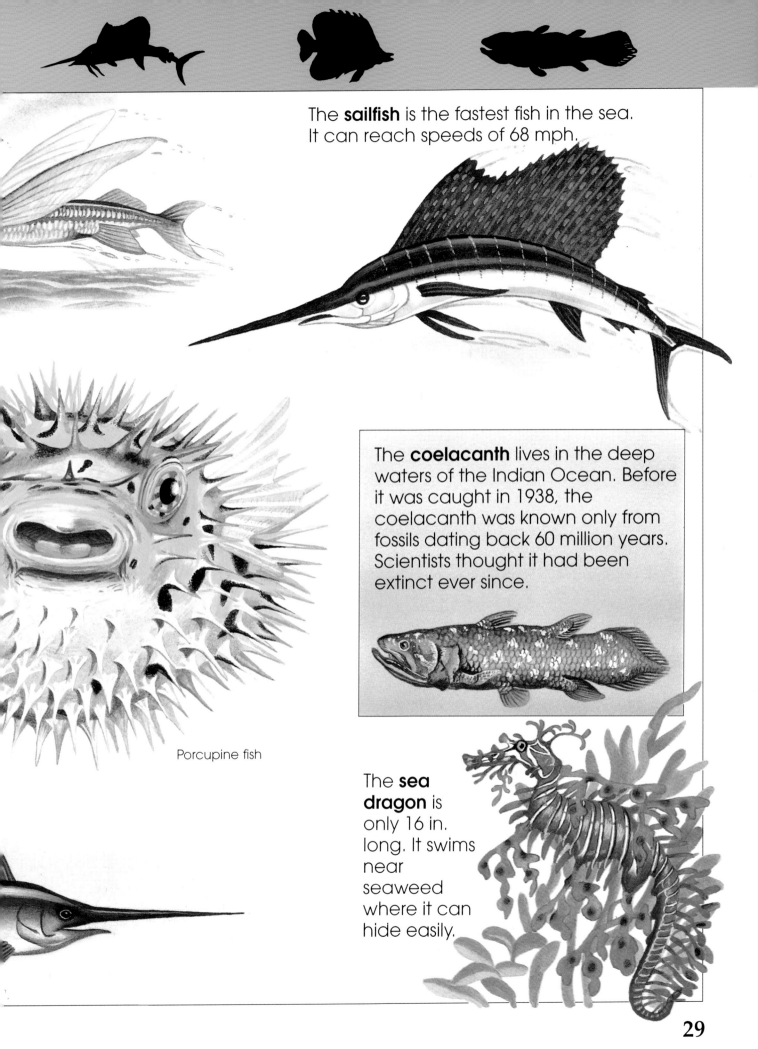

The **sailfish** is the fastest fish in the sea. It can reach speeds of 68 mph.

Porcupine fish

The **coelacanth** lives in the deep waters of the Indian Ocean. Before it was caught in 1938, the coelacanth was known only from fossils dating back 60 million years. Scientists thought it had been extinct ever since.

The **sea dragon** is only 16 in. long. It swims near seaweed where it can hide easily.

Millions of years ago, strange creatures lived in the oceans.

Scientists know about these creatures because they have found fossils of their bones buried in ancient rocks.

Many of these giant sea animals lived at the same time as the dinosaurs.

Kronosaurus had the largest head of any hunter in the sea. It was almost 10 ft. long and was armed with lots of sharp teeth.

Cryptocleidus had strong flippers to propel it through the water. It caught small fish in its long jaws armed with sharp teeth.

Ichthyosaurus looked like a dolphin or large fish, but was really a reptile. Ichthyosaurus could not come on shore to lay eggs like most reptiles, so it gave birth to live young.

Archelon was the largest turtle. It was nearly 13 feet long and lived about 70 million years ago.

Nothosaurus
was one of the first reptiles to live in the sea. It lived in Europe about 210 million years ago.

Tanystropheus lived on the coast. It dipped its long neck in to the water and snapped up fish and shrimp.

Placodus lived about 200 million years ago in Europe. It ate shellfish and used its webbed feet to help it swim.

Placodus

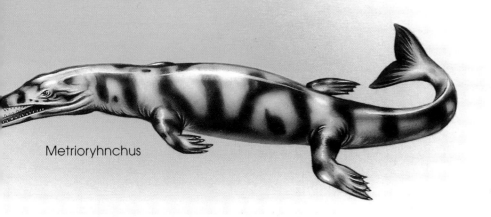

Metrioryhnchus

Metriorhynchus was a 10-foot long crocodile which lived in the ocean 140 million years ago. It was so adapted to life in the sea that it even had a fish-like tail.

Because the oceans are so vast, there are many areas which have never been explored properly.

Sailors who have traveled off the main shipping routes have reported seeing strange and curious creatures. As nobody has ever caught one of these mysterious creatures, scientists do not believe they really exist.

The type of **sea monster** most often seen has a small head and a long neck held upright. Witnesses say they see a large body under the water with four large fins which move the creature slowly along.

This sea monster looks like a prehistoric sea animal, **elamosaurus**, which was about 33 feet long.

The 13-foot long **megamouth shark** was not discovered until the 1980s. Nobody knew about it until one was accidentally caught in a net. This proved that large sea creatures can exist without anybody knowing about them.

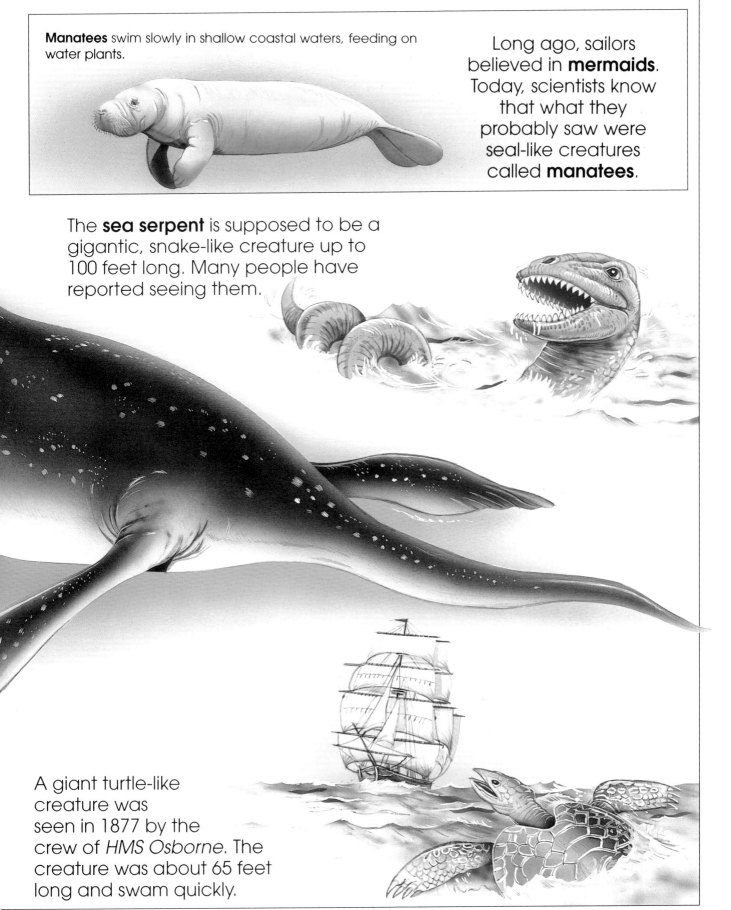

Manatees swim slowly in shallow coastal waters, feeding on water plants.

Long ago, sailors believed in **mermaids**. Today, scientists know that what they probably saw were seal-like creatures called **manatees**.

The **sea serpent** is supposed to be a gigantic, snake-like creature up to 100 feet long. Many people have reported seeing them.

A giant turtle-like creature was seen in 1877 by the crew of *HMS Osborne*. The creature was about 65 feet long and swam quickly.

Many young animals look like small versions of their parents. They simply grow into adults. Others look completely different. They develop into adults through stages.

Some young creatures that look like small adults, such as snails, grow very gradually. Others, such as insects, have a hard outer skeleton. They have to molt in order to grow.

They molt by making a new, soft skeleton beneath the hard one. The new skeleton is pumped up with air, and this splits the old skeleton. The young creature grows inside the new, hard skeleton.

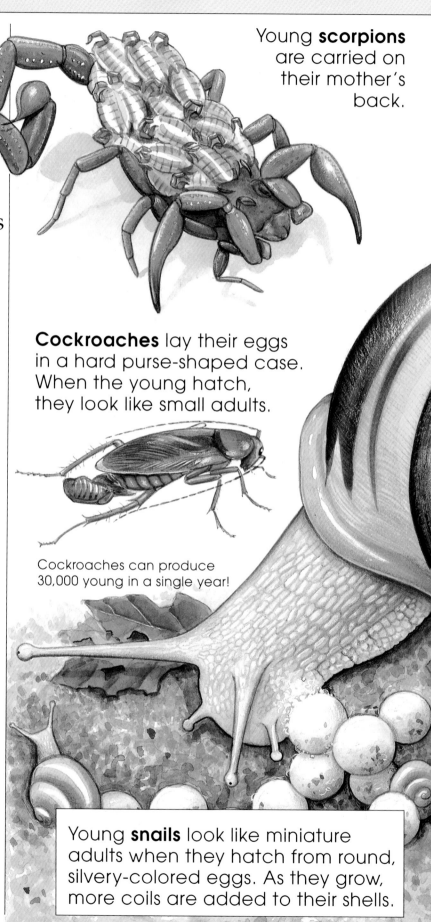

Young **scorpions** are carried on their mother's back.

Cockroaches lay their eggs in a hard purse-shaped case. When the young hatch, they look like small adults.

Cockroaches can produce 30,000 young in a single year!

Young **snails** look like miniature adults when they hatch from round, silvery-colored eggs. As they grow, more coils are added to their shells.

34

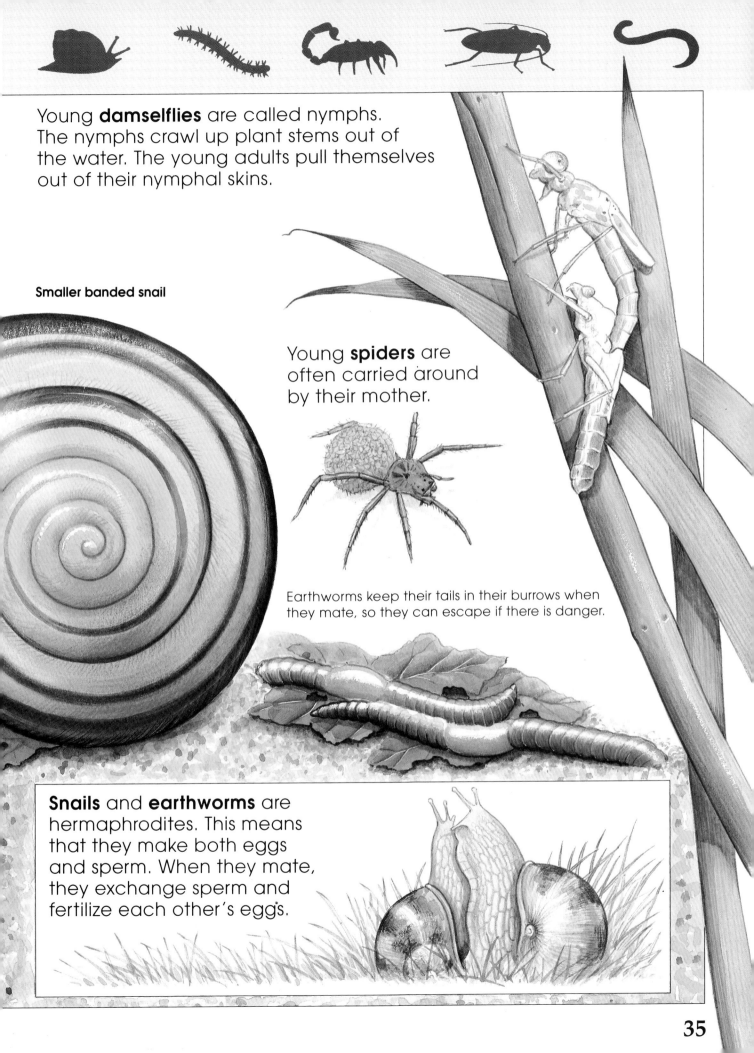

Young **damselflies** are called nymphs. The nymphs crawl up plant stems out of the water. The young adults pull themselves out of their nymphal skins.

Smaller banded snail

Young **spiders** are often carried around by their mother.

Earthworms keep their tails in their burrows when they mate, so they can escape if there is danger.

Snails and **earthworms** are hermaphrodites. This means that they make both eggs and sperm. When they mate, they exchange sperm and fertilize each other's eggs.

When they are born, many young creatures look completely different from their parents. They go through several stages to develop into adults.

The young that hatch from the eggs are called larvae. A larva feeds and grows. It eventually develops into a chrysalis, which is also called a pupa. Inside the chrysalis, the larva changes into an adult. After a time, the adult emerges from the chrysalis.

The development of a larva into an adult through these stages is called **metamorphosis.**

Lady bug beetles and their larvae feed on aphids.

Caddis fly larvae live underwater. They make homes to live in by sticking together pieces of plant, sand, shells and other material. They carry their homes around with them.

The larvae of **butterflies** and **moths** develop into adults through metamorphosis.

This egg has been laid by a **pasha** butterfly.

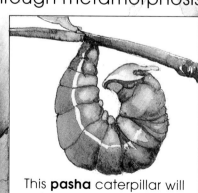

This **pasha** caterpillar will turn into a chrysalis.

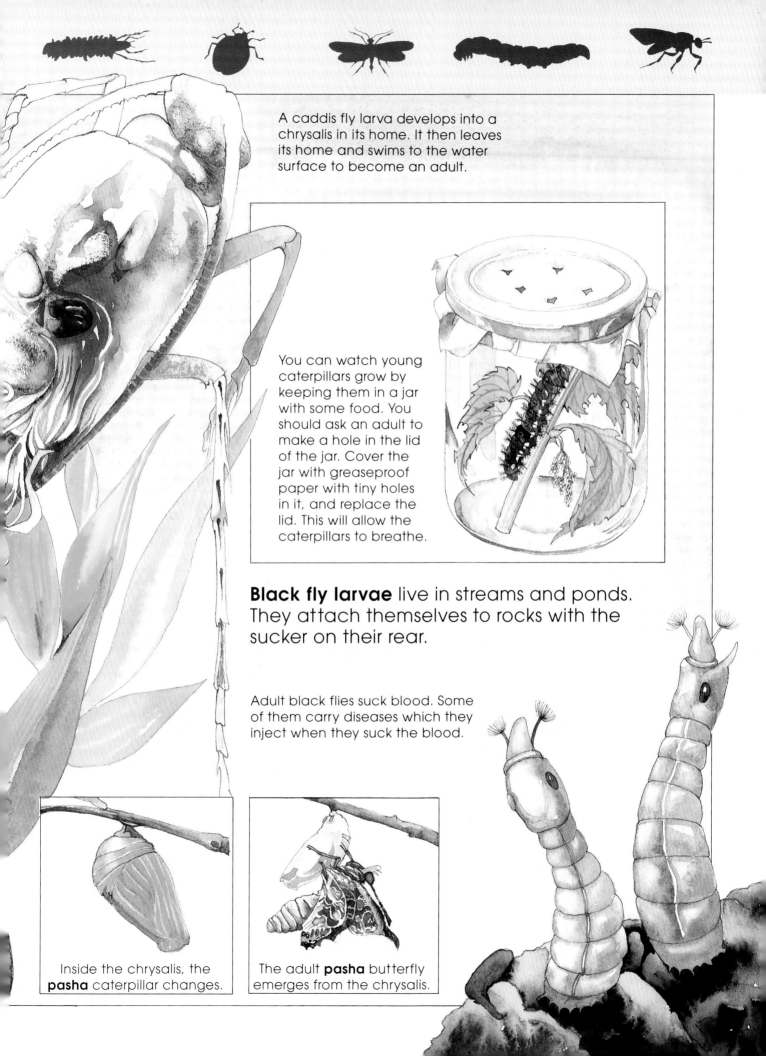

A caddis fly larva develops into a chrysalis in its home. It then leaves its home and swims to the water surface to become an adult.

You can watch young caterpillars grow by keeping them in a jar with some food. You should ask an adult to make a hole in the lid of the jar. Cover the jar with greaseproof paper with tiny holes in it, and replace the lid. This will allow the caterpillars to breathe.

Black fly larvae live in streams and ponds. They attach themselves to rocks with the sucker on their rear.

Adult black flies suck blood. Some of them carry diseases which they inject when they suck the blood.

Inside the chrysalis, the **pasha** caterpillar changes.

The adult **pasha** butterfly emerges from the chrysalis.

Hunters and trappers

Small creatures have to find food to eat in order to grow. Some of them eat the leaves, shoots, flowers, fruits and roots of plants. Many creatures even eat other creatures!

Little creatures find their food in different ways. Some of them eat rotten plants or animals, while others suck juices from plants, or even blood from animals!

Some small creatures tunnel and burrow through the soil, while others hunt for a meal on the surface of the ground. Some minibeasts even make traps in which they catch their prey.

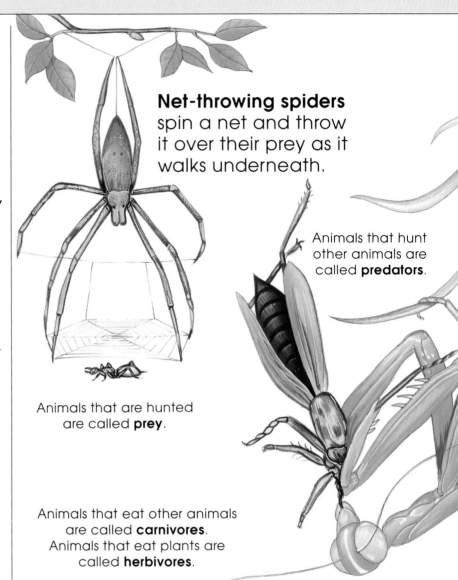

Net-throwing spiders spin a net and throw it over their prey as it walks underneath.

Animals that hunt other animals are called **predators**.

Animals that are hunted are called **prey**.

Animals that eat other animals are called **carnivores**. Animals that eat plants are called **herbivores**.

Long-jawed spiders are well camouflaged on grass as they wait for their prey to walk past.

Trap-door spiders hide in a silk-lined burrow with a trap door at the entrance. They throw open the trap door to grasp their prey.

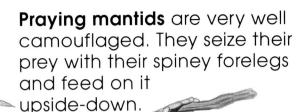

Praying mantids are very well camouflaged. They seize their prey with their spiney forelegs and feed on it upside-down.

Antlion larvae lie half-hidden at the bottom of a funnel-shaped pit. They flick sand at creatures that slip over the edge of the pit, so that the tiny creatures fall down to the bottom.

Blue-black spider wasps have a loud buzz which terrifies their prey.

Euglandina rosea attacking a **papustyla** snail.

Tiger beetles are fierce hunters. They use their strong jaws to kill and cut up their prey, which includes young lizards.

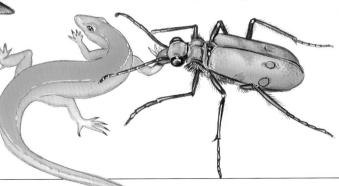

Snails sometimes attack and eat other snails. If the snail has withdrawn inside its shell, the attacker will drill a hole through the shell to eat the snail.

Rotters, tunnellers and burrowers

Nothing lives for ever. Plants and animals die and rot away. Many feed on rotting material, helping to break it down into smaller particles.

Some of these pass into the soil, and are eaten by burrowing creatures. The particles contain nutrients which help them to grow.

Mole crickets dig burrows with their large spade-like feet. They eat the roots of plants and other insects.

Scarab beetles make balls of dung, and bury them in a tunnel where they lay their eggs. The larvae discover a pantry full of lovely food.

Mites help to break down the remains of dead plants in the soil. Some of them feed on fungi, while others hunt other mites.

Dermistid beetles help to tidy up the remains of dead animals.

Many different kinds of small creatures can be found in compost heaps. To find them, place a handful of compost in a sieve and warm it gently with a lamp for two to three days. Remember to keep the tissue paper damp.

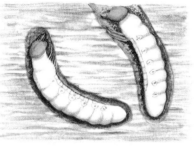

Stag beetle larvae live inside logs. They tunnel through the decaying wood for several years before they emerge as adult stag beetles.

Termites tunnel into wood or soil and build nests. These hang from trees, or are huge mounds rising from the ground.

Microscopic animals live inside termites. They break down the plant food that termites eat.

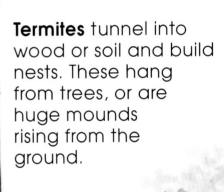

Millipedes tunnel through the soil. They eat particles of rotten material which are rich in nutrients. They also eat fallen leaves, breaking them down into smaller pieces.

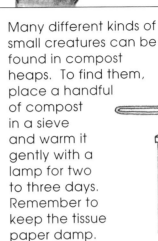

41

Suckers

Some insects live on liquid food. They have extremely sharp mouthparts which they use to pierce the skin of an animal or the tissue of a plant.

They usually suck blood or plant juices through a sucking tube.

Fleas use the hooks and spines on their bodies to hold tightly on to the fur or skin of their hosts. Fleas can carry diseases which they inject into their hosts when they bite.

In the Middle Ages, the disease called the Black Death was spread by the rat flea. This disease killed millions of people.

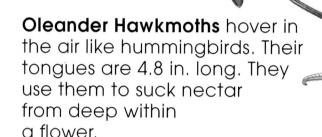

Oleander Hawkmoths hover in the air like hummingbirds. Their tongues are 4.8 in. long. They use them to suck nectar from deep within a flower.

Ticks are parasites. They sink their hooked mouthparts into the flesh of their host. As they suck the blood, their round, elastic bodies swell greatly.

Aphids feed on plant juices. Their delicate mouthparts pierce the sap vessels inside a plant, and the pressure forces sweet-tasting sap into the aphid's body.

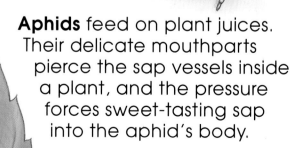

Some of the sap is passed out of the aphid as a drop of sweet fluid. This is sometimes eaten by ants.

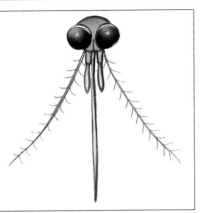

The long, needle-sharp mouthparts of a mosquito contain a sucking tube.

Mosquitoes feed on blood and plant juices. Female mosquitoes have a meal of blood before they lay their eggs. Male mosquitoes suck plant juices instead of blood.

Female mosquitoes bite humans. A person can lose over a pint of blood in an hour.

Jungle leeches suck blood. When they have had a blood meal, their body swells.

Animals that live and feed on other animals are called **parasites**. The animals that provide a home and food are called **hosts**.

Thrips are tiny "thunder-bugs." They have mouthpieces on one side of their mouth only, which they use to suck plant juices.

Thrips are pests, feeding on corn and other crops.

Robber flies catch and stab their prey with their sharp mouthpieces. Their victim is then sucked dry.

Cochineal bugs suck plant juices. They are used to make food coloring as they are dark red.

Flyers

Small creatures have to get around in order to find food, a mate, and a new place to live. They also need to be able to escape from predators.

Insects use many ways of getting around. Some of them crawl and others run. Some of them jump and others wriggle. Some of them can even fly.

Most insects that fly have two pairs of wings which beat together.

Spiders are creatures that can fly but do not have wings!

A young **wolf spider** has released a long, silken thread. The wind will pluck the thread into the air, whisking the young spider away with it.

Beetles have two pairs of wings. The first pair is very tough and protects the delicate flying wings which are folded underneath when not in use.

Dragonflies chase other flying insects by rapidly beating their outstretched wings.

Emperor dragonfly

Damselflies fly by fluttering their wings. They catch other flying insects by grasping them with their legs.

Butterflies fly during the daytime. Most of them slowly flap their large, colorful wings.

The wings of the **Painted lady** warn other insects to keep away.

The wings of the **swallowtail** make a noise as they clap together.

Flies are the best acrobats of the insect world. They can even land upside-down on a ceiling.

Flies have only one pair of real wings. The rear wings are tiny bat-shaped objects which beat very fast.

Hover flies can hover, dart backward and forward, and even fly straight upward.

Midges have one of the fastest wing beats. Some beat their wings over 1,000 times a second.

Fairy flies have delicate, feathery wings. They are one of the smallest flying insects.

Cockchafers fly at dusk. They can fly over 3 miles in search of a mate.

Crawlers and runners

Many small creatures get around by crawling or running. Some of them have lots of short legs which they use to crawl about.

Other creatures have fewer legs, but they are usually quite long. Long legs allow the creature to run about quickly.

Pseudoscorpions can run backward as well as forward! They are active hunters that crawl among decaying leaves in search of a meal.

Pseudoscorpions have long sensitive hairs on their rear to help them feel where they are going.

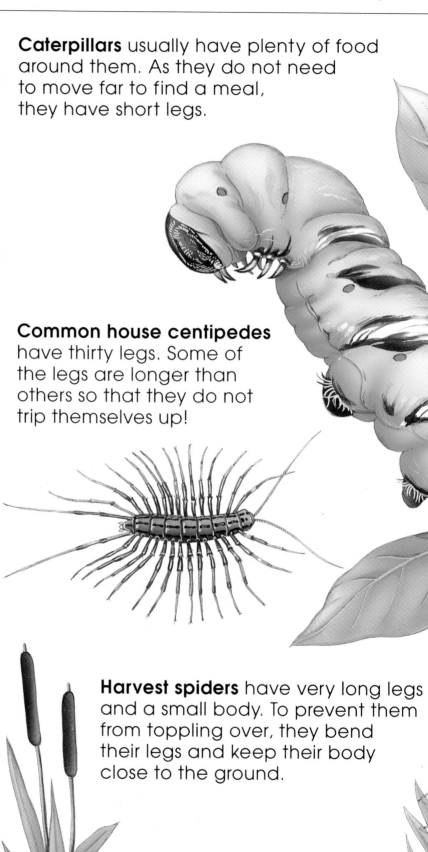

Caterpillars usually have plenty of food around them. As they do not need to move far to find a meal, they have short legs.

Common house centipedes have thirty legs. Some of the legs are longer than others so that they do not trip themselves up!

Harvest spiders have very long legs and a small body. To prevent them from toppling over, they bend their legs and keep their body close to the ground.

The legs of harvest spiders are not used for speed. The spiders crawl through the vegetation where they live.

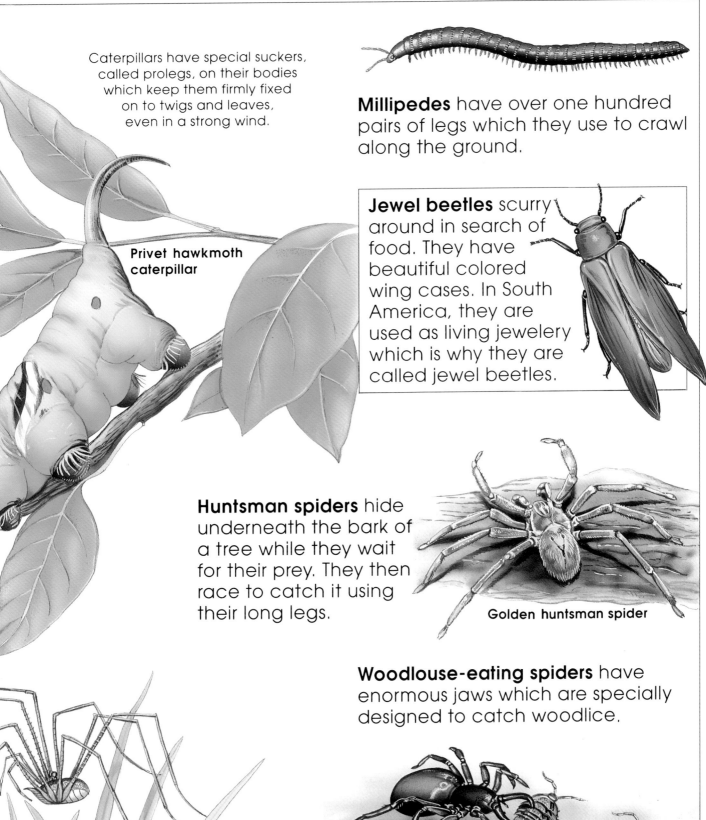

Caterpillars have special suckers, called prolegs, on their bodies which keep them firmly fixed on to twigs and leaves, even in a strong wind.

Millipedes have over one hundred pairs of legs which they use to crawl along the ground.

Privet hawkmoth caterpillar

Jewel beetles scurry around in search of food. They have beautiful colored wing cases. In South America, they are used as living jewelery which is why they are called jewel beetles.

Huntsman spiders hide underneath the bark of a tree while they wait for their prey. They then race to catch it using their long legs.

Golden huntsman spider

Woodlouse-eating spiders have enormous jaws which are specially designed to catch woodlice.

All spiders have eight legs, which they use to run about.

Hoppers, jumpers and skaters

Some small creatures move around by hopping and jumping. Being able to jump suddenly is a good way to catch a meal, or to escape from a predator.

Some insects skate across the surface of water in search of food or a mate.

Raft spiders stand half on the water and half on a water plant. They race across the water surface to catch their prey, which includes small fish.

Grasshoppers and **crickets** have huge back legs. They use the strong muscles in these legs to catapult themselves high into the air.

Fleas have large back legs which allow them to jump very high - well over half a yard.

Fleas jump onto animals, such as cats, where they make their home.

Treehoppers hop from tree to tree in search of food.

Pond skaters have waterproof hairs on their feet which help them to float on the water surface.

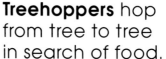

Grasshoppers attract a mate by rubbing their back legs against their front wings to make a singing sound.

Springtails can spring suddenly into the air using their special "tail."

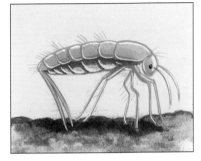

The "tail" is tucked under the springtail's body.

The "tail" straightens suddenly, making the springtail spring into the air.

Jumping plant lice have very strong back legs which means they can jump from plant to plant.

Apple suckers are jumping plant lice which live on apple trees.

Click beetles have a peg on their bodies. When they lie on their backs and bend, the peg pops free with a loud click, and they jump into the air.

Whirligig beetles skate quickly across the surface of a pond in a zigzag pattern.

Jumping spiders have excellent sight. When they see a fly, they will leap into the air to catch it.

Slitherers and wrigglers

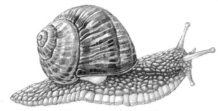

Legs can get in the way, so some creatures do not have any legs at all. They have soft bodies, and they move around by slithering along the ground or wriggling through the soil.

Earthworms make burrows which let air into the soil. They drag leaves into the burrows for food.

Leeches move along by using their suckers. They have two suckers on their bodies, one at the front and one at the rear. The one at the front has teeth as it is also their mouth.

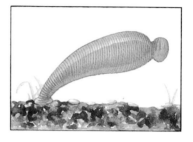

The rear sucker sticks to the ground and the body stretches forward.

The front sucker sticks to the ground and the body is pulled forward.

African giant snail

You can make a wormery by putting some earthworms and compost into a plastic bucket with small holes in the bottom. As the earthworms eat the compost, you will need to add some more to the bucket.

Earthworms burrow through the soil by eating it. They grip the soil with very small bristles along their bodies.

Slugs and snails are special creatures that slither along on a trail of slime using one foot. If you place a slug or snail on a piece of clear plastic and look at it from underneath, you will see ripples moving along the foot as the slug moves forward.

Hover fly larvae look like little leeches. They wriggle along in search of aphids which they eat.

Fly larvae hatch from eggs laid on dung. They have small legs, or no legs at all. To move about, they wriggle through their squidgy food.

Soil centipedes have up to 100 pairs of tiny legs which help them to grip the soil.

Nematodes are minute roundworms which live inside many animals and plants, and in soil. They move around by wriggling their tiny bodies.

51

Tricksters

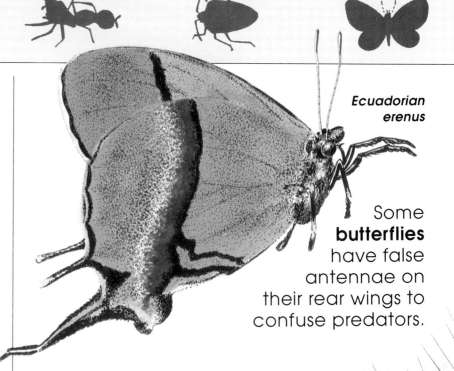

Ecuadorian erenus

Many insects try to trick predators by using different disguises. Some of them have the same colors as creatures that are fierce or poisonous, so that predators will leave them alone.

Some **butterflies** have false antennae on their rear wings to confuse predators.

Other insects let predators approach them, but then they give them a nasty surprise. A few even use false heads to confuse their predators!

Tussock moth caterpillars have fine irritating hairs on their body which give predators a nasty shock!

Diadem butterflies are not poisonous but they trick their predators by flying with poisonous **African monarch** and **Citrus swallowtail butterflies**.

Citrus swallowtail

African monarch

Wasp beetles are not dangerous as they do not sting. They pretend to be wasps to trick their predators.

Copying the color of another creature is called **mimicry**. This helps to protect harmless insects from predators.

Some **jumping spiders** mimic mutillid wasps to protect themselves.

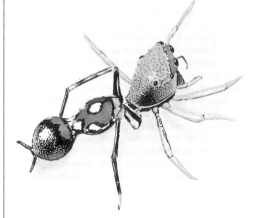

The jumping spider's rear looks like the head of a mutillid wasp.

Shieldbugs ooze a stinking liquid when they are in danger. This is why they are also called stinkbugs.

Golden-silk spiders have bright colors. At a distance, these break up the shape of the spider, making it difficult to see.

Diadem

Flashers and warners

Many creatures use bright colors to protect themselves. Some of them frighten their predators by suddenly flashing bright colors at them.

Some insects show their bright colors all the time. Predators learn that these are warning colors, telling them that the insect is dangerous.

There are only a few warning colors: black, white, yellow, red and brown. Creatures learn quickly that these colors warn of danger.

Peacock butterflies have large colorful eye spots on their wings.

Puss moth caterpillars shoot out long red tassles from tubes on their rear when they are frightened.

South African grasshoppers have brightly colored bodies which warn other creatures to keep away.

Owl butterflies become fierce creatures with large, staring eyes when their wings are spread open.

Owl butterflies look like leaves with their wings folded.

Cotton stainer bugs are left alone by birds because of their warning colors.

Praying mantis rise up and startle their predators with a brilliant pattern of colors.

Lantern bugs have colorful wings. They warn predators to keep away.

Singers and glowers

Many creatures use sound to attract a mate, or to warn off predators. Some of them make sounds during the day. If you walk through a field or a forest, you may hear all kinds of chirps and buzzes.

Many insects make sounds at night, while others use light to attract a mate. The males or females glow in the dark, and their mates are attracted to them.

Katydids sing their repetitive song "katydid, katydidn't" at night. They sing by rubbing their left front wing against a ridge on the right wing.

Katydids and other crickets have ears on their legs.

Tree crickets make thousands of piercing chirps without stopping. Some tree crickets can be heard 1 mi. away!

58

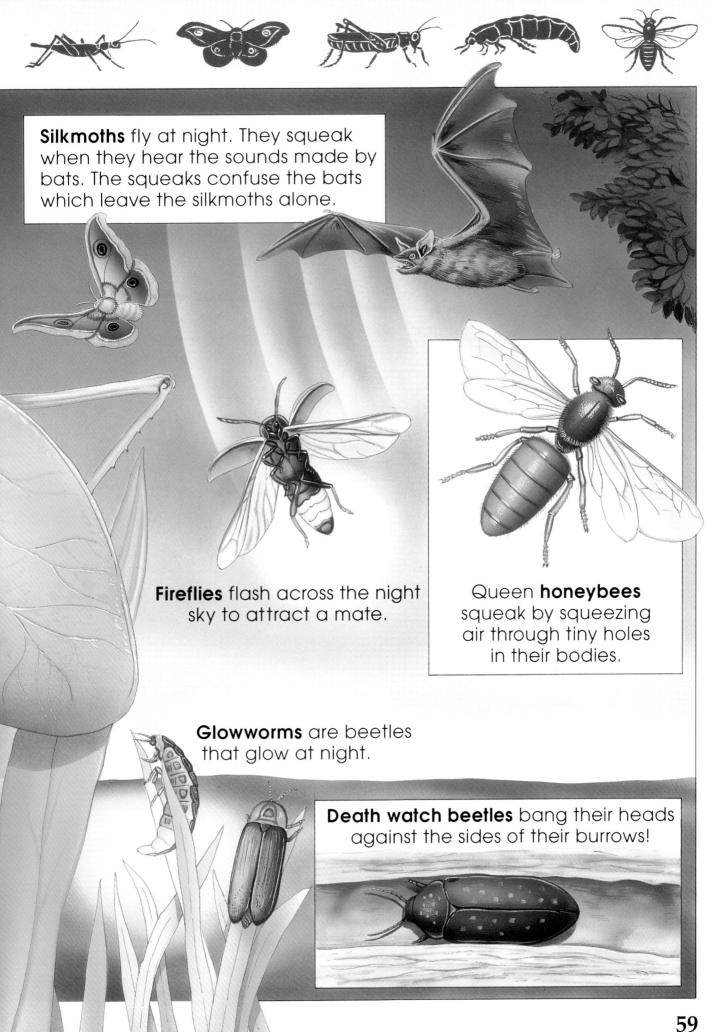

Silkmoths fly at night. They squeak when they hear the sounds made by bats. The squeaks confuse the bats which leave the silkmoths alone.

Fireflies flash across the night sky to attract a mate.

Queen **honeybees** squeak by squeezing air through tiny holes in their bodies.

Glowworms are beetles that glow at night.

Death watch beetles bang their heads against the sides of their burrows!

Most creatures leave their young to look after themselves. Many of the young starve to death, or are eaten by predators. To overcome this, many eggs are laid.

Some creatures care for their eggs and young, so fewer eggs need to be laid. Insects such as female ants or bees work together to provide shelter and food for their young, giving them a better chance of survival.

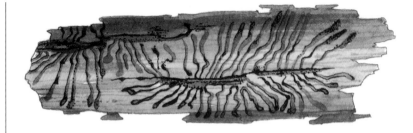

Look at the bark of fallen trees and see if you can find the tunnels of bark beetles.

Elm bark beetles tunnel under the bark of a tree where they lay their eggs.

Termites live as a family in a huge nest. The king and queen live in the royal chamber. The queen's body swells to a huge size as she lays her eggs inside it. She can lay 30,000 eggs a day.

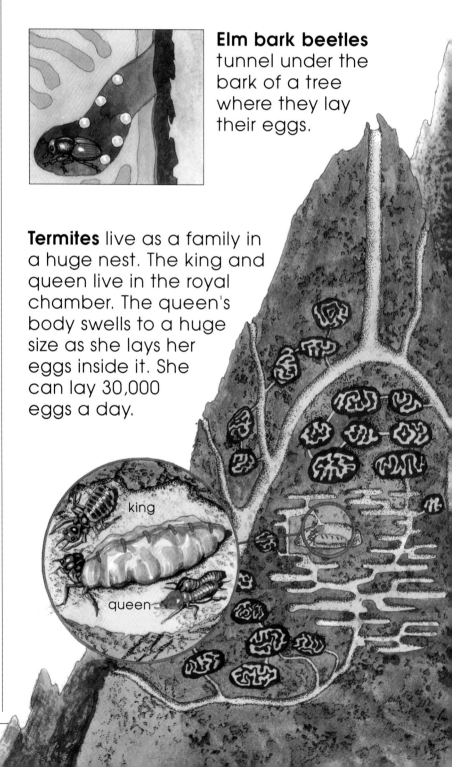

king

queen

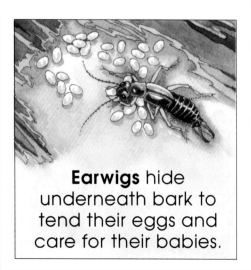

Earwigs hide underneath bark to tend their eggs and care for their babies.

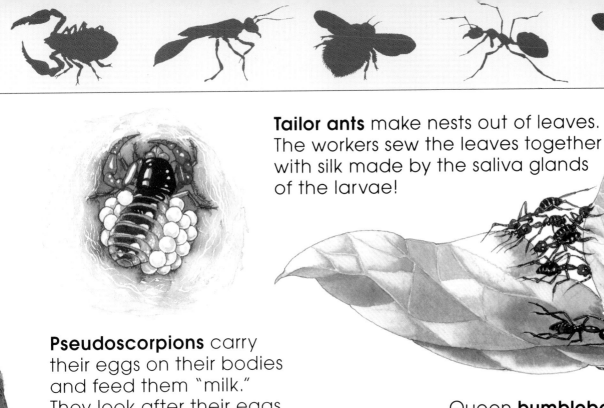

Tailor ants make nests out of leaves. The workers sew the leaves together with silk made by the saliva glands of the larvae!

Pseudoscorpions carry their eggs on their bodies and feed them "milk." They look after their eggs and young inside a tiny nest made of silk.

Queen **bumblebees** build a wax honeypot in the nest where they lay their eggs, so they have plenty of food. They care for the young on their own.

Sandwasps catch and sting a caterpillar. This sends it to sleep. They put it in a burrow in the sand, and lay an egg on it.

When the egg hatches, the larva feeds on the sleeping caterpillar.

Oak gall wasps lay their eggs on the rib of an oak leaf. The rib swells and forms a gall which is a safe home for the larvae that grow inside it.

Galls come in all shapes and sizes. The aleppo gall is used to make special permanent ink which is used by banks.

61

Monsters on land

Happily there are few really monstrous large animals. Smaller monsters are much more common.

Many large animals might look frightening, but usually they do not attack unless they are threatened.

Pythons swallow their prey whole.

Giant **pythons** coil their powerful bodies around their helpless prey until they suffocate it.

The **elephant** is the largest land mammal. A full grown male (bull) African elephant can be over 10 ft. tall and weigh 9 tons.

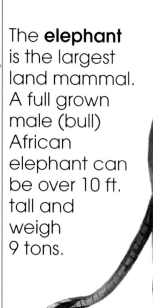

The enormous **Komodo Dragon** prowls through the forest on lonely Indonesian islands.

This fierce lizard, nearly three metres long, will even attack people.

A **tiger** has huge, sharp teeth which grip and kill its prey.

The **giraffe** is the tallest mammal on Earth. However, it is not fierce and eats only leaves.

Grizzly bears tower a frightening 10 ft. when they stand upright on their hind legs. They have big, sharp claws for tearing at food.

Gorillas are the largest primates. When threatened, a male gorilla will beat his chest with his hands, roar and rush toward the enemy.

The **Goliath beetle** is a heavy weight champion of the insect world. It can carry a load 850 times its own weight. That is similar to a human carrying 67 tons.

Some of the strangest monsters can be found swimming and living in the sea.

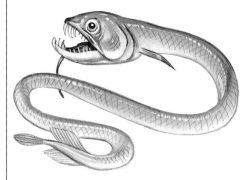

Large ones like the whales and sharks swim in the open ocean. Others, like giant sponges, hide deep down on the seabed.

Lurking at the bottom of the sea near Japan are **giant spider crabs**. With their claws outstretched they can measure nearly ten feet.

The suckers on a 50 foot **giant squid** measure 4 in. across. But sucker scars on whales have been seen as long as 18 in.!

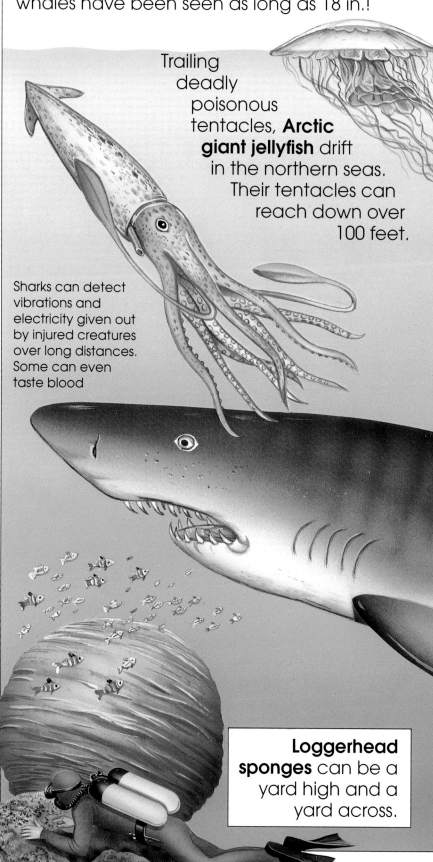

Trailing deadly poisonous tentacles, **Arctic giant jellyfish** drift in the northern seas. Their tentacles can reach down over 100 feet.

Sharks can detect vibrations and electricity given out by injured creatures over long distances. Some can even taste blood

Loggerhead sponges can be a yard high and a yard across.

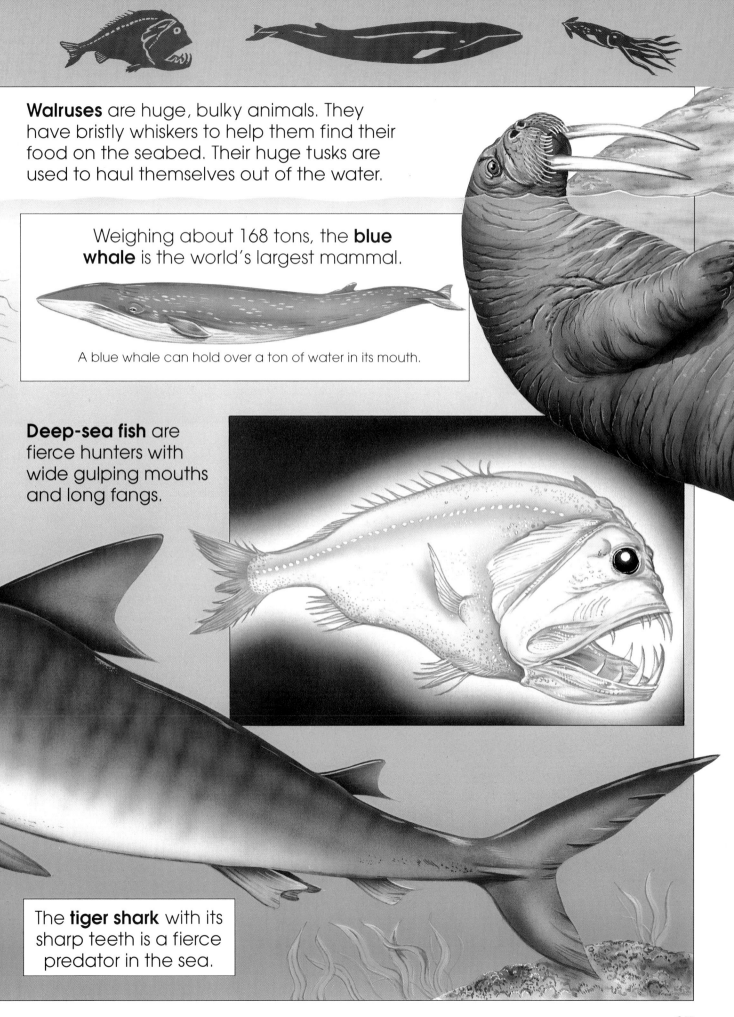

Walruses are huge, bulky animals. They have bristly whiskers to help them find their food on the seabed. Their huge tusks are used to haul themselves out of the water.

Weighing about 168 tons, the **blue whale** is the world's largest mammal.

A blue whale can hold over a ton of water in its mouth.

Deep-sea fish are fierce hunters with wide gulping mouths and long fangs.

The **tiger shark** with its sharp teeth is a fierce predator in the sea.

Monsters in the air

Birds, bats and insects all have wings and can fly. Some are fierce hunters in the air and can grow very large.

Others use their long, needle-sharp claws, called talons, to catch and kill.

Monstrous **robber·flies** hunt other insects in the air, piercing them with sharp mouthpieces, and sucking out the contents of their bodies.

Bats are the only mammals that can truly fly. The largest bat is the **flying fox** which can have a wingspan of over 6 feet.

The **Andean condor** is the world's largest bird of prey and can weigh over 25 lbs.

Albatrosses circle the Earth, only coming to land when they want to breed.

The wings of an **albatross** can span more than 10 feet. They enable it to fly hundreds of miles at a time.

The wingspan of the largest moth in the world, the **atlas moth**, is 10 in.

Pelicans dive to catch fish. Nearly half a yard long, their huge bills scoop up several fish at a time which they then swallow.

This evil-looking **wasp** has paralyzed another insect with its terrible sting.

Monster animals covered with hair can look very strange. They are hairy for many reasons.

Some live in very cold places and need to keep warm. Others use hair for camouflage.

Poisonous hairs protect against attack. Hairs are even used to help some animals breathe underwater.

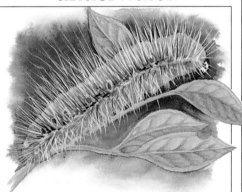

The hairs on this **Japanese Dictyoploca moth caterpillar** irritate and hurt any predator trying to eat it.

The body of a **porcupine** is covered with special hairs. When frightened the animal rattles these needle-sharp quills.

Some porcupines can even shoot quills out at their enemy.

The "old man of the forest," or **orang-utang**, has very long, golden red hair.

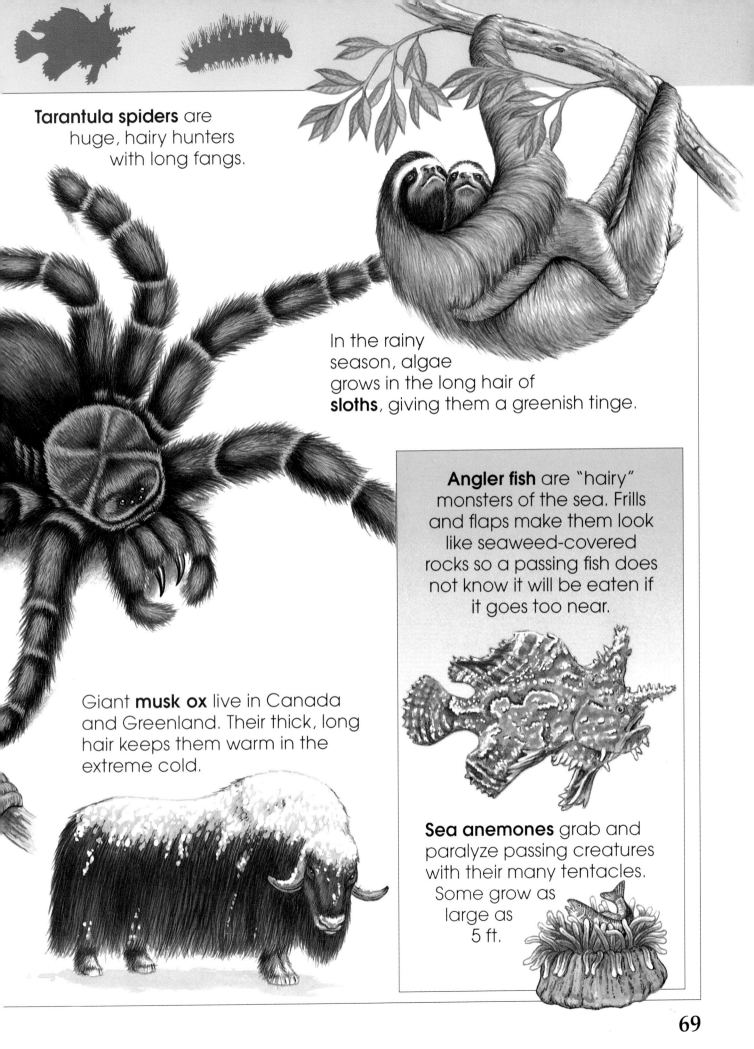

Tarantula spiders are huge, hairy hunters with long fangs.

In the rainy season, algae grows in the long hair of **sloths**, giving them a greenish tinge.

Angler fish are "hairy" monsters of the sea. Frills and flaps make them look like seaweed-covered rocks so a passing fish does not know it will be eaten if it goes too near.

Giant **musk ox** live in Canada and Greenland. Their thick, long hair keeps them warm in the extreme cold.

Sea anemones grab and paralyze passing creatures with their many tentacles. Some grow as large as 5 ft.

To protect themselves from being attacked or eaten, many animals are monstrous looking.

Some look frightening all the time, while others can make themselves scary when they have to.

Roaring and puffing up their bodies are just some of the methods used.

Death's head hawkmoths can enter beehives and steal honey without being stung.

The strange skull-like markings on the **death's head hawkmoth** give it a deathly appearance.

This is not a fierce prehistoric monster, but a **frilled lizard**. This harmless lizard puts on an impressive display when it is frightened.

Male **stag beetles** have huge, fearsome jaws. They cannot bite with them, but instead joust with other males over females.

Stag beetles use their huge jaws to try and flick their opponent over.

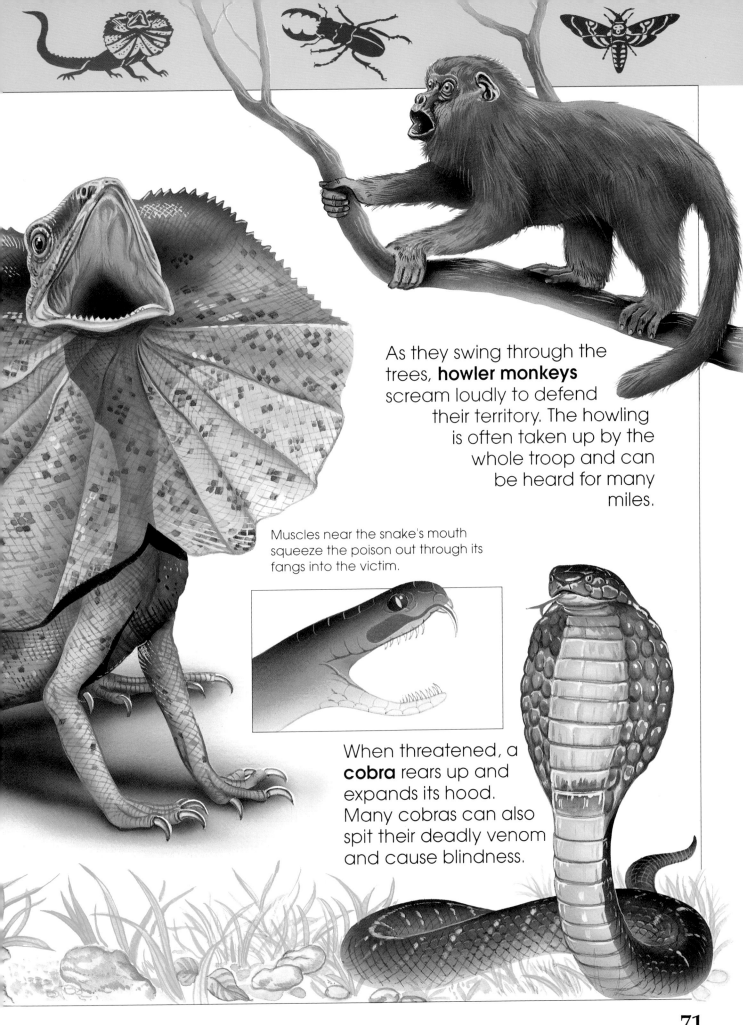

As they swing through the trees, **howler monkeys** scream loudly to defend their territory. The howling is often taken up by the whole troop and can be heard for many miles.

Muscles near the snake's mouth squeeze the poison out through its fangs into the victim.

When threatened, a **cobra** rears up and expands its hood. Many cobras can also spit their deadly venom and cause blindness.

Disgusting monsters

Some monster animals use horrid smells to frighten their predators.

Others live in smelly places or have disgusting habits.

Eating dung and rotting corpses is not particularly nice, but without these animals to clear up, the world would be even smellier!

Big dung beetles carefully roll dung into balls which they hide in tunnels underground for their grubs to eat.

When a **vampire bat** finds a sleeping animal, it bites into the skin with its razor-sharp front teeth and laps up the blood with its tongue.

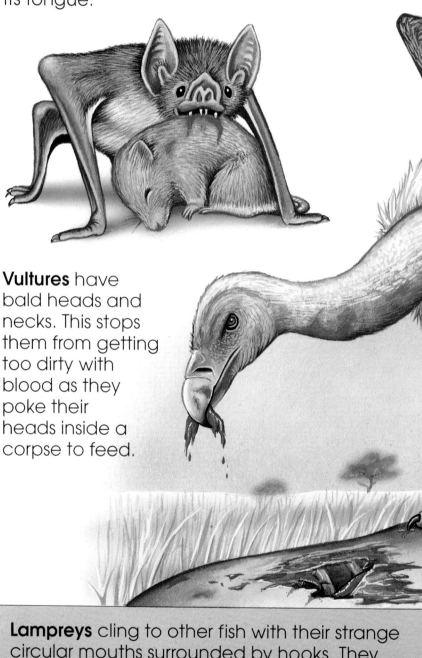

Vultures have bald heads and necks. This stops them from getting too dirty with blood as they poke their heads inside a corpse to feed.

Lampreys cling to other fish with their strange circular mouths surrounded by hooks. They gnaw the flesh and even wriggle into their host's body.

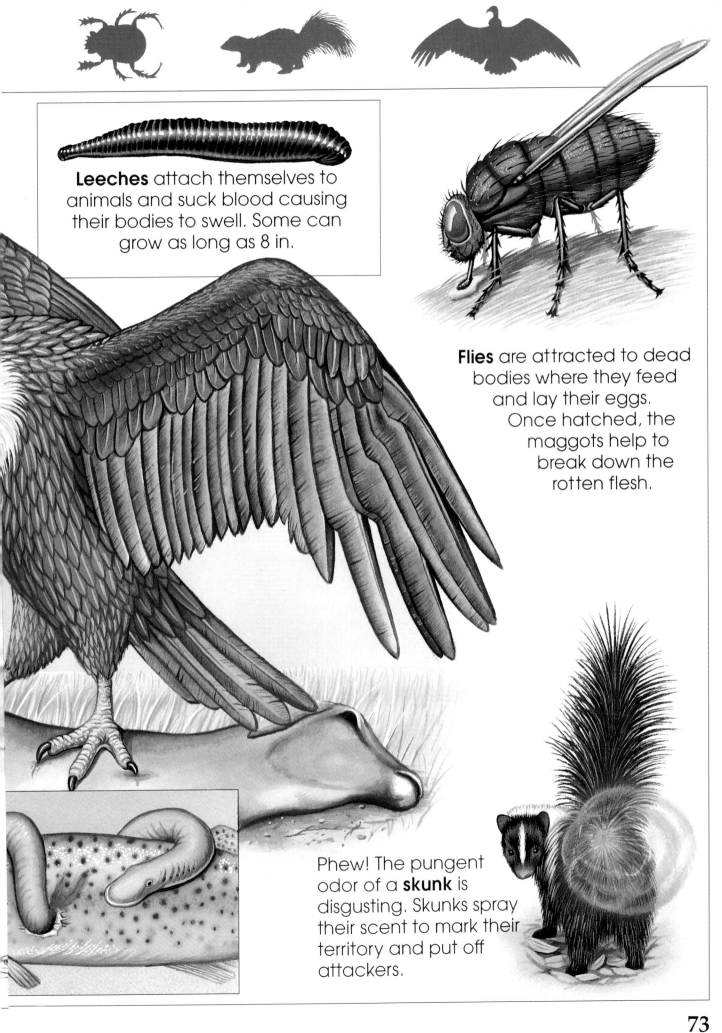

Leeches attach themselves to animals and suck blood causing their bodies to swell. Some can grow as long as 8 in.

Flies are attracted to dead bodies where they feed and lay their eggs. Once hatched, the maggots help to break down the rotten flesh.

Phew! The pungent odor of a **skunk** is disgusting. Skunks spray their scent to mark their territory and put off attackers.

Fat monsters

Some animals are monstrously fat. Many of them spend most of their time in the water where the weight of their bodies is supported.

Fat bodies can hold a lot of food for times when there is little food around. They can also be used to scare off attackers.

The fat **Vietnamese pot-bellied pig** is kept as a pet in some parts of the world.

Herds of **elephant seals** wallow on the beach. An adult male can weigh almost four tons. When males fight each other they often crush the babies on the beach.

Hippopotamus means "river horse." Although they look fat and clumsy on land, when they are in water they can swim fast.

Hippos use their large teeth for digging up water plants and fighting.

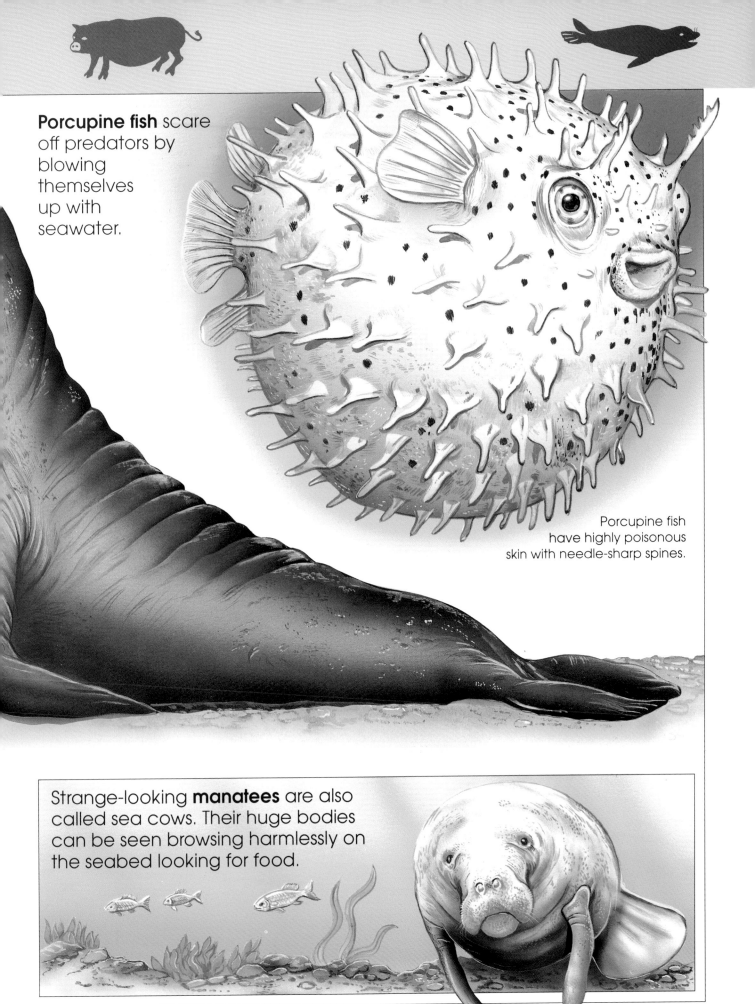

Porcupine fish scare off predators by blowing themselves up with seawater.

Porcupine fish have highly poisonous skin with needle-sharp spines.

Strange-looking **manatees** are also called sea cows. Their huge bodies can be seen browsing harmlessly on the seabed looking for food.

Weird monsters

Some animals are very strange-looking to us. But these monsters are usually the shape they are for a reason.

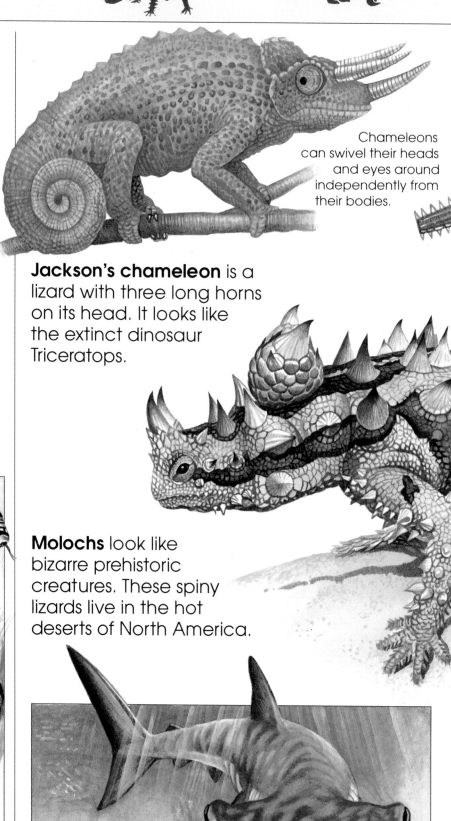

Chameleons can swivel their heads and eyes around independently from their bodies.

Everything is made so that it is suited to where it lives so it can survive.

Jackson's chameleon is a lizard with three long horns on its head. It looks like the extinct dinosaur Triceratops.

Molochs look like bizarre prehistoric creatures. These spiny lizards live in the hot deserts of North America.

Animals do not usually have two heads, but sometimes they are born. This freak two-headed **kingsnake** was found in California.

The strange-looking **hammerhead shark** is a ferocious hunter. It even attacks people.

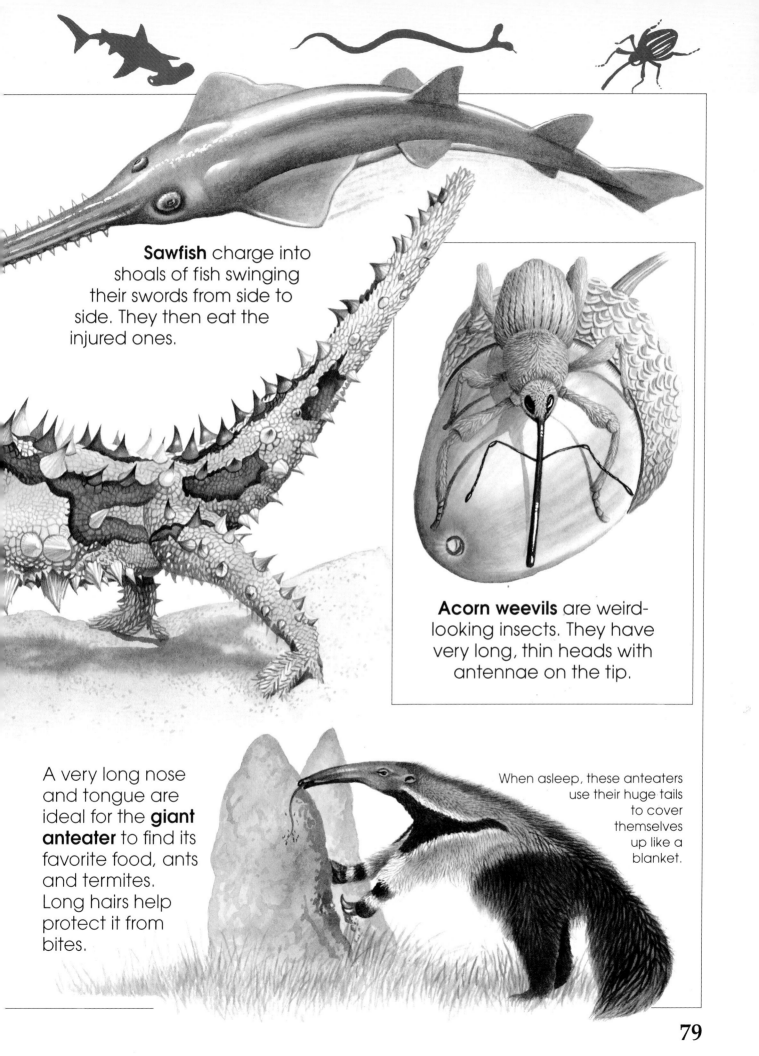

Sawfish charge into shoals of fish swinging their swords from side to side. They then eat the injured ones.

Acorn weevils are weird-looking insects. They have very long, thin heads with antennae on the tip.

A very long nose and tongue are ideal for the **giant anteater** to find its favorite food, ants and termites. Long hairs help protect it from bites.

When asleep, these anteaters use their huge tails to cover themselves up like a blanket.

Deadly monsters

Many animals protect themselves from attack by stinging or biting.

Some animals use poison to stun or kill their prey. Many of these deadly animals have ways of warning others to keep away!

A **black widow spider** traps its prey in a web and then sucks out its insides.

Long, brightly colored spines cover the body of the beautiful but deadly **lion fish**. The sharp spines are coated with toxic mucus and cause terrible pain if touched.

The long trailing tentacles of the **Portuguese Man O'War jellyfish** are highly poisonous. Stinging cells shoot tiny barbed harpoons into anything that touches them.

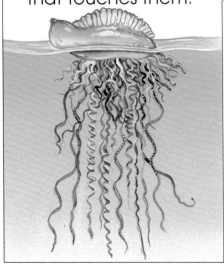

The bright colors of **poison dart frogs** warn predators to leave them alone.

People living in the rainforests of South America smear their blow-pipe darts with the frog's mucus (slime) to poison their prey.

Many **sea urchins** are covered in sharp, poisonous spines for protection. If stepped on the spines can stab and break off in your foot.

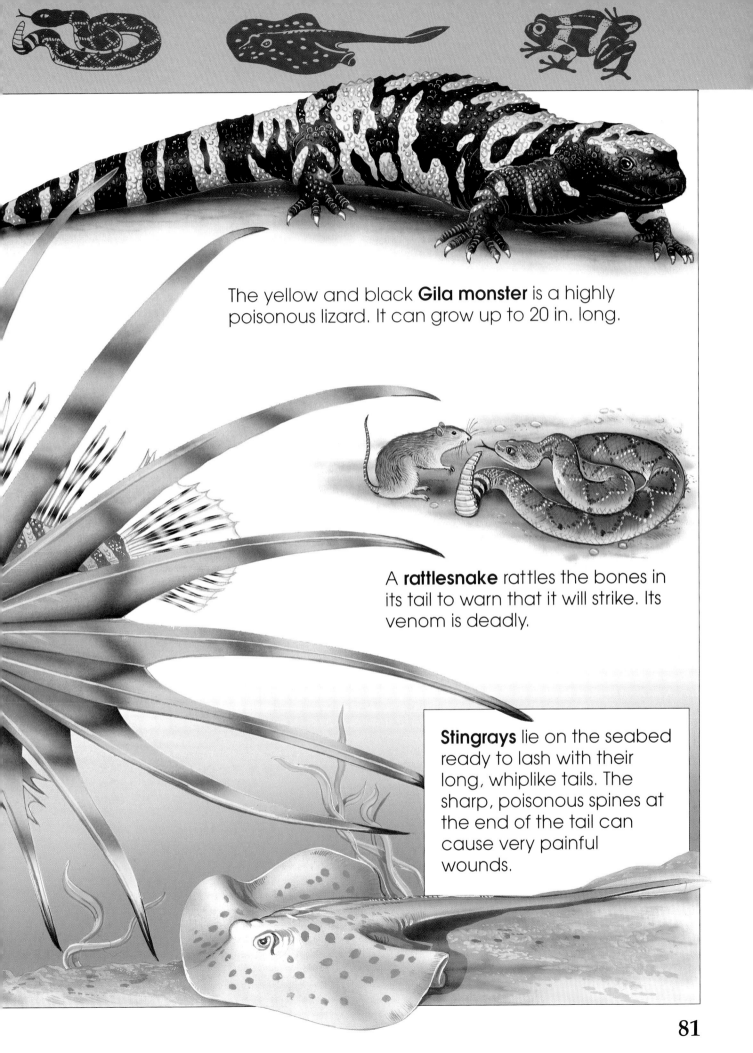

The yellow and black **Gila monster** is a highly poisonous lizard. It can grow up to 20 in. long.

A **rattlesnake** rattles the bones in its tail to warn that it will strike. Its venom is deadly.

Stingrays lie on the seabed ready to lash with their long, whiplike tails. The sharp, poisonous spines at the end of the tail can cause very painful wounds.

Masses of monsters

Some animals are only frightening and dangerous in large numbers.

Some, like bees, live together in groups to help each other. Others, like wolves, hunt in packs.

Some animals only group together in masses at certain times.

In some parts of the world, plagues of flying **locusts** can darken the sky, eating every green plant they land on.

Hornets live as a colony, nesting inside hollow trees. They use their huge jaws and deadly sting to hunt.

A colony of **army ants** marching through the forest will eat everything in its way - even small animals.

82

Millions of **mosquitoes** often breed together. The females must have a blood meal before they can lay their eggs. A person can lose nearly a pint of blood to these insects if they are not protected.

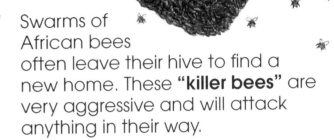

Swarms of African bees often leave their hive to find a new home. These **"killer bees"** are very aggressive and will attack anything in their way.

African wild dogs live in packs of up to 60. By circling their prey and dashing in and biting it, the victim is soon weakened and killed.

Hunting together in packs, **wolves** can catch and kill large animals. They usually attack the weak and sick, but rarely people.

Rare monsters

Many animals are becoming rare. Some have already become extinct and will never be seen again outside a museum.

People kill animals for their skin, fur, feathers and horns. We also destroy the places where they live.

The largest **false scorpion** in Europe lives under the bark of dead trees. It is now extremely rare and only found in ancient forests.

On Maria Island in the West Indies lives the world's rarest snake, the **St. Lucia racer**. There are less than 100 left.

Racers are large, fast snakes which strike repeatedly with their heads when attacked, tearing the flesh.

Wildlife parks and zoos do important work trying to save animals from extinction. The last wild **Californian condor** was captured so it could breed under protection.

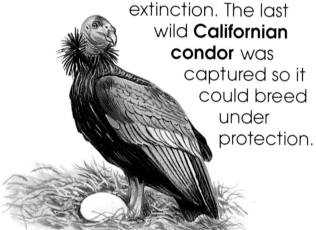

Trap door spiders in Southeast Asia are the rarest spiders. They use their jaws to dig holes, leaving a hinged lid at the entrance. When a victim comes near, the spider opens the lid, grabs its prey, and pulls it underground.

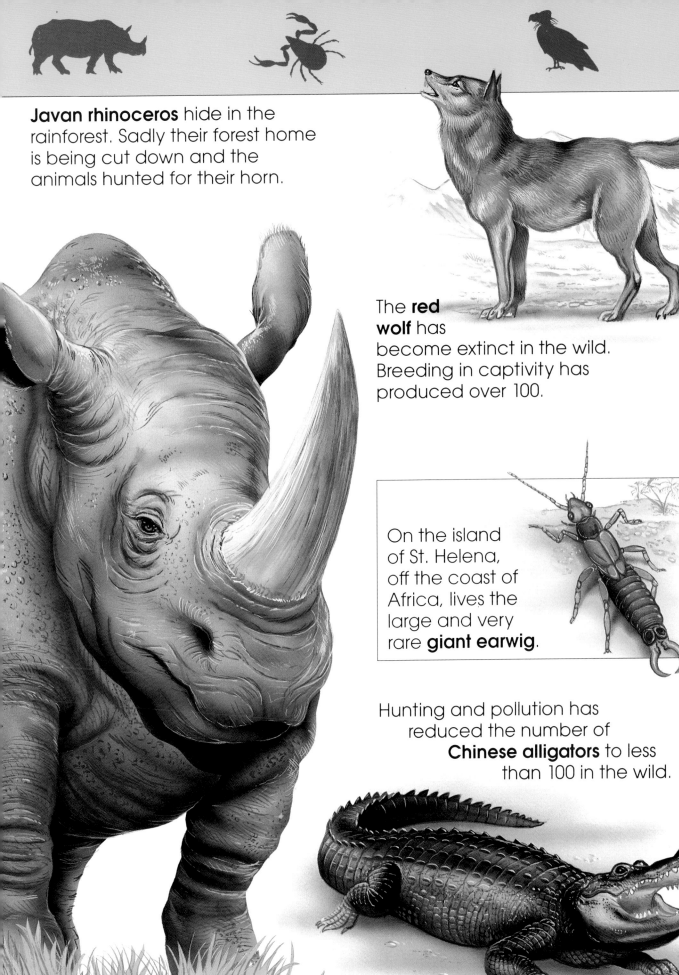

Javan rhinoceros hide in the rainforest. Sadly their forest home is being cut down and the animals hunted for their horn.

The **red wolf** has become extinct in the wild. Breeding in captivity has produced over 100.

On the island of St. Helena, off the coast of Africa, lives the large and very rare **giant earwig**.

Hunting and pollution has reduced the number of **Chinese alligators** to less than 100 in the wild.

Imaginary monsters

Superstition and fear have made people dream up all kinds of strange and imaginary monsters.

Some of these unnatural creatures were invented from stories of unusual animals brought back by travelers.

Other mythical monsters are based on actual living, and extinct, animals.

Some imaginary monsters might be real, we just do not know for sure.

Every year, thousands of people watch the water on Loch Ness in Scotland, hoping to see the **Loch Ness Monster**. Some believe that the monster could be a surviving plesiosaur, a prehistoric sea creature.

The **hydra** is a nine-headed beast of Greek mythology. It was very difficult to kill as each time a head was cut off it grew two new ones.

Sailors' sightings of mermaids may have been **dugongs**. At a distance, female dugongs with their young look like women holding their babies in their arms.

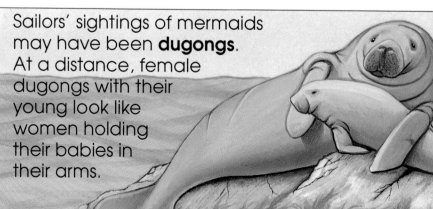

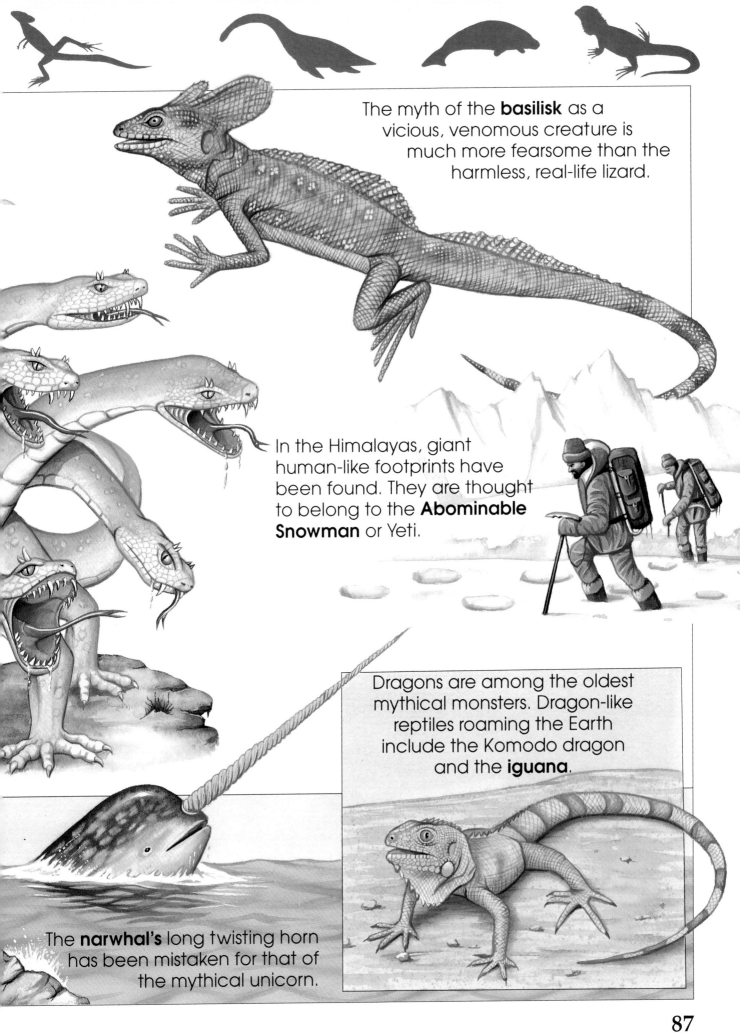

The myth of the **basilisk** as a vicious, venomous creature is much more fearsome than the harmless, real-life lizard.

In the Himalayas, giant human-like footprints have been found. They are thought to belong to the **Abominable Snowman** or Yeti.

Dragons are among the oldest mythical monsters. Dragon-like reptiles roaming the Earth include the Komodo dragon and the **iguana**.

The **narwhal's** long twisting horn has been mistaken for that of the mythical unicorn.

Millions of years ago, all kinds of strange monstrous animals roamed the Earth.

There were no people around when the dinosaurs ruled the world.

When people appeared they cut down forests and hunted animals. Some of the larger species were driven into extinction. Today, people still kill and threaten many animals.

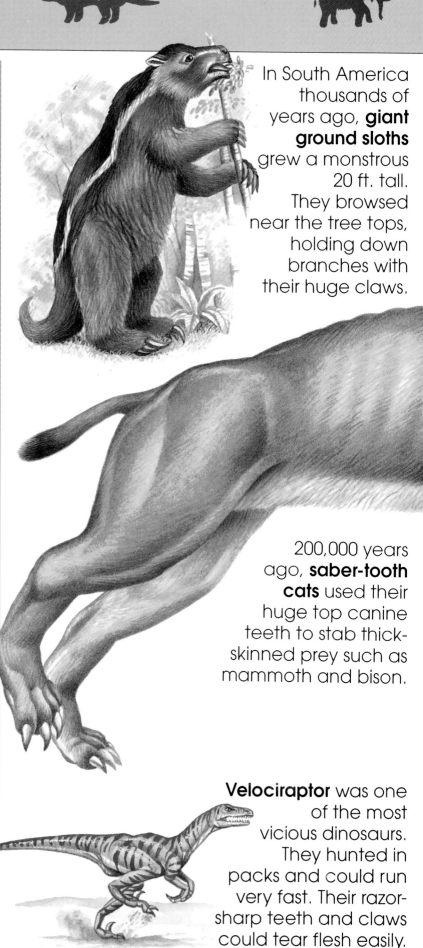

In South America thousands of years ago, **giant ground sloths** grew a monstrous 20 ft. tall. They browsed near the tree tops, holding down branches with their huge claws.

200,000 years ago, **saber-tooth cats** used their huge top canine teeth to stab thick-skinned prey such as mammoth and bison.

Velociraptor was one of the most vicious dinosaurs. They hunted in packs and could run very fast. Their razor-sharp teeth and claws could tear flesh easily.

Quetzalcoatlus' wings were made of skin like those of bats today.

97 million years ago, **quetzalcoatlus** soared through the air on wings spanning 40 feet.

Giant Irish deer grew antlers nearly 13 feet across. They died out 2,500 years ago.

Mammoths are one of the largest land mammals to have lived. They grew over 13 ft. tall and had woolly coats and huge tusks.

The giant **moa** of New Zealand was the tallest bird that ever existed. It stood over 10 feet tall.

People destroyed the moa's habitat and hunted it, so that by 1800 it was extinct.

Hyaenodon must have been a fearsome hunter and scavenger. Its skull was 26 in. long and full of needle-sharp teeth.

89

The Japanese eat fugu fish. But if the wrong part of the fish is eaten by mistake, it is poisonous enough to kill.

Is there danger under the sea?

In Australia, a swimmer was stung by a jellyfish. He died just 30 seconds later!

Seas cover four-fifths of the world. In places, the sea is deeper than the highest mountain, Mount Everest, which is 29,028 feet high. Many strange creatures live in the sea, and some of them are very dangerous indeed.

Shark

Q Do any fish eat humans?

A Several types of shark will attack and eat humans. The most famous is the great white shark, which can grow to be over 20 feet long. It can swallow a human whole! There are about 50 shark attacks on people each year, but because there is often no trace of the victims, the total may be much higher.

Q Do killer whales deserve their name?

A Yes. Killer whales (orcas), are the most dangerous of the toothed whales, and even attack whales which are twice as big as they are. They hunt in large packs and use many cunning tactics. They leap on to beaches to kill seals. Sometimes they find penguins and seals on ice floes. The whales push the ice floes to knock them into the water. Killer whales even smash their way through the ice to reach their prey.

Q Which is the most poisonous fish?

A The most poisonous fish in the world is the stonefish. When it lies on the bottom of shallow tropical seas, it looks exactly like a stone. If you step on it, the stonefish injects poison into your foot. The sting is incredibly painful. Most people die from the sting unless they receive treatment quickly.

Stonefish

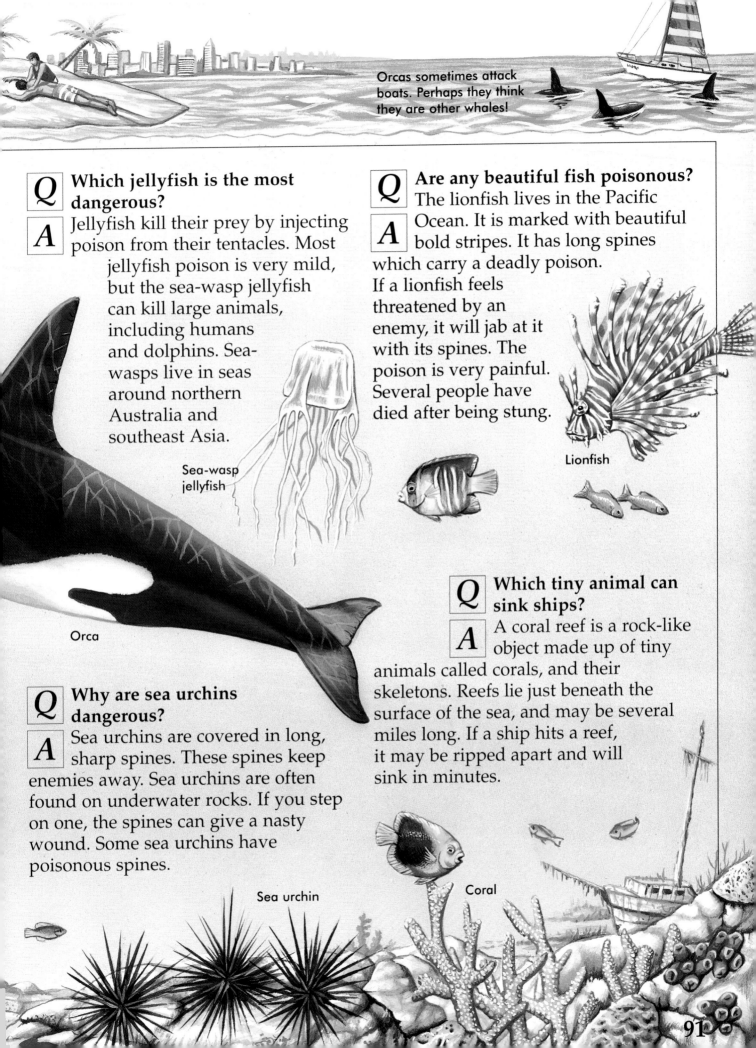

Orcas sometimes attack boats. Perhaps they think they are other whales!

Q Which jellyfish is the most dangerous?

A Jellyfish kill their prey by injecting poison from their tentacles. Most jellyfish poison is very mild, but the sea-wasp jellyfish can kill large animals, including humans and dolphins. Sea-wasps live in seas around northern Australia and southeast Asia.

Sea-wasp jellyfish

Orca

Q Why are sea urchins dangerous?

A Sea urchins are covered in long, sharp spines. These spines keep enemies away. Sea urchins are often found on underwater rocks. If you step on one, the spines can give a nasty wound. Some sea urchins have poisonous spines.

Sea urchin

Q Are any beautiful fish poisonous?

A The lionfish lives in the Pacific Ocean. It is marked with beautiful bold stripes. It has long spines which carry a deadly poison. If a lionfish feels threatened by an enemy, it will jab at it with its spines. The poison is very painful. Several people have died after being stung.

Lionfish

Q Which tiny animal can sink ships?

A A coral reef is a rock-like object made up of tiny animals called corals, and their skeletons. Reefs lie just beneath the surface of the sea, and may be several miles long. If a ship hits a reef, it may be ripped apart and will sink in minutes.

Coral

The fire salamander has a poisonous skin. People used to believe that this protected it from flames.

Are any frogs poisonous?

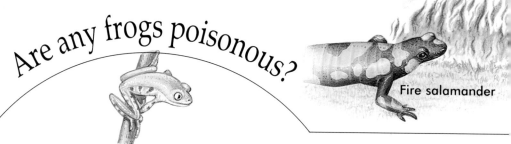

Fire salamander

Frogs are amphibians – cold-blooded creatures which usually live on land and breed in water. Amphibians secrete an unpleasant-tasting liquid from their skin, which helps to keep enemies away. In some types of frog, this liquid is poisonous.

Q How do humans use frog poison?

A The most poisonous frogs live in South and Central America. Some peoples collect the frogs and bake them to remove the poison from the skin. The concentrated poison is then smeared on arrows used for hunting and warfare. A single scratch from such an arrow can kill.

Q Which is the most poisonous frog?

A The golden arrow-poison frog may be the most dangerous frog in the world. It is about 1.5 inches long. If a small animal eats a golden arrow-poison frog, the poison will kill it. A larger animal will become very sick.

Q Is the common toad poisonous?

A The common toad is found across Europe, northern Asia and North Africa. It usually hides from danger, but if it feels threatened, it produces a foul-tasting poisonous liquid.

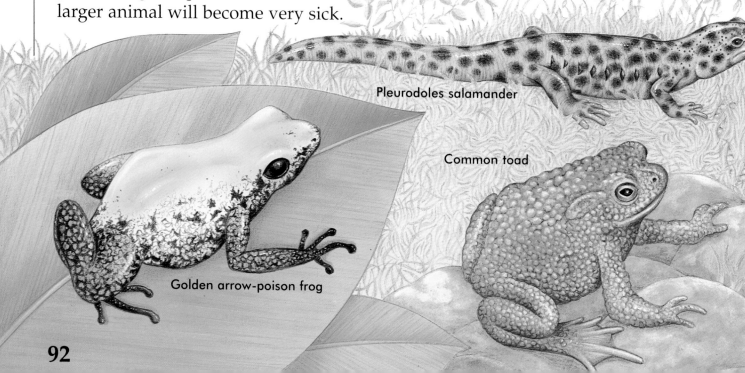

Pleurodoles salamander

Common toad

Golden arrow-poison frog

92

The first person to report the Komodo dragon to scientists was a policeman who chased a suspect to the island of Komodo in 1910.

Komodo dragon

Q Are any lizards poisonous?

A The gila monster is a lizard found in Mexico and Arizona. It can deliver a poisonous bite. The poison is produced in its lower jaw and emerges through holes near its teeth. The gila monster chews its victim to work the poison into the wound. If a human is attacked, he can usually shake the lizard off before much poison is absorbed.

Q Which toad frightens its enemies?

A The fire-bellied toad's back is colored like a normal toad, but its underside is bright red with black spots. When it is attacked, the fire-bellied toad flashes its red underside to warn that it has a deadly poisonous skin.

Q What is the Komodo dragon?

A Komodo dragons live on the island of Komodo, near Java. They are enormous lizards which can grow to over 10 feet in length, maybe twice as long. They are ferocious hunters and prey on deer, pigs, and other animals.

Q Which salamander is spiky?

A The pleurodoles salamander lives in southern Europe and North Africa. It is born with long ribs, which have very sharp ends. When the salamander has grown into an adult, the ribs poke through its skin to form sharp spikes, often surrounded by red patches of skin.

Komodo dragon

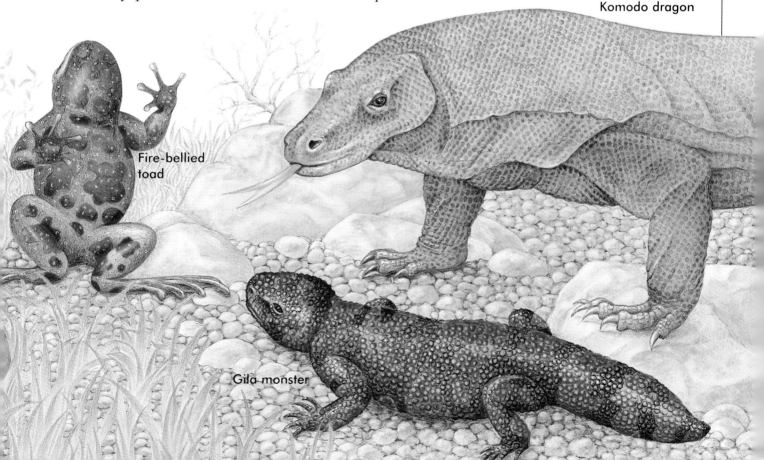

Fire-bellied toad

Gila monster

Are insects dangerous?

At Peshawar in Pakistan, one scorpion stung nine people, killing eight of them.

In the 1340s, a plague spread by rat fleas killed one-third of the people in Europe.

There are about a million different types of insect. Many feed on plants, but some are hunters which prey on other insects and small animals. Many insects spread deadly diseases. Some spiders have a painful bite. A bite from a scorpion can kill.

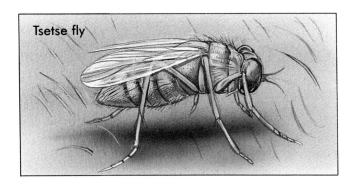

Tsetse fly

Q Which diseases are spread by insects?

A Insect bites can transmit many diseases. One of the most dangerous is sleeping sickness, which is spread by the tsetse fly. In parts of Africa where there are lots of tsetse flies, it is not safe for people to live there. Some mosquitoes spread malaria, a fever which sometimes kills.

Q Why are houseflies dangerous?

A Houseflies are found all over the world. Although they do not sting or bite, they can be dangerous. They feed both on fresh food, and on rotting food or manure. A fly landing on your meal may bring with it germs and dirt from its previous meal.

Q Which insects suck blood?

A Mosquitoes and some flies bite humans to suck their blood. The blackfly is common on sandy beaches in North America and can give a very painful bite. Mosquito bites often make your skin come up in a red lump, which may become very itchy.

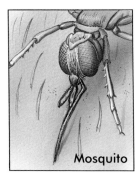

Mosquito

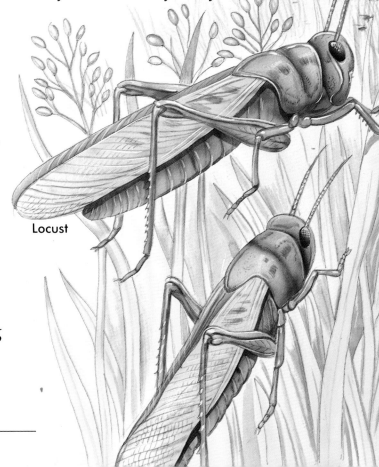

Locust

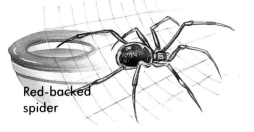

Red-backed spider

The red-backed spider of Australia likes dark, damp places and often hides in bathrooms. It can give you a painful bite!

Q Which insects can destroy crops?

A Many different types of insect eat crops, but locusts may cause spectacular damage. Each locust can eat its own weight in food in a day. A large swarm of locusts can destroy the crops of an entire country, causing a famine which kills many people. In 1889, a huge swarm containing 250,000 billion locusts covered 1,875 square miles in Egypt.

Q Why do Africans shake their boots?

A Scorpions are active at night. When dawn comes, they like to hide in a dark, warm place. Sometimes they crawl into boots or shoes. People in Africa usually shake their boots before putting them on in the morning, in case they contain a deadly scorpion. A scorpion sting can kill.

Q Which insects eat each other?

A Mantids are insects which hunt other insects. They sit completely still on a leaf or flower, and then grab passing prey. Mantids are so aggressive that they will eat each other, even their own mates.

Q What is found in bee stings?

A The stings of bees and wasps can be very dangerous. They contain chemicals that cause a sharp pain. Some people are especially sensitive to the venom, and may become very ill if they are stung.

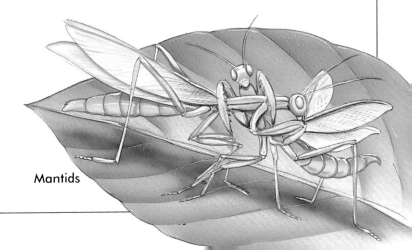

Mantids

The Australian taipan has enough poison to kill over 200,000 mice, its main prey.

Are all snakes deadly?

Taipan

There are 2,700 species of snake and every one of them hunts other animals for food. However, fewer than 200 species are known to have killed any people.

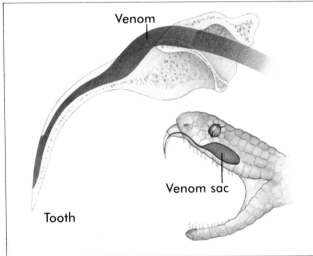

Venom

Venom sac

Tooth

Q Are all snakes poisonous?

A About 300 species of snake are venomous. There are about 100 species which kill their prey by constriction. This means the snake coils itself around its victim, slowly squeezing until the victim cannot breathe, and suffocates.

Boa constrictor

Q How do snakes inject poison?

A Some types of snake use poison to kill their prey. The poison, or venom, is produced and stored in a gland inside the skull, and is connected to the teeth. They have special hollow teeth which inject poison like a hypodermic needle. When a snake bites its prey, the poison runs into the wound and enters the animal's bloodstream.

Q How many people are killed by snakes each year?

A Snakes will attack a human only if they feel threatened or have been surprised, perhaps by being trodden on in long grass. It is thought that about 30,000 people die each year after being bitten by snakes.

The Indian cobra can spit poison to blind an enemy up to two metres away.

Cobra

Four times more men than women are bitten by snakes, because men spend more time working in fields.

Anaconda

Q Which is the world's largest snake?

A The largest snake in the world is the anaconda of South America, which is known to grow to about 30 feet in length. But in 1907, explorer Percy Fawcett shot an anaconda which he thought was even longer – about 60 feet long. Unfortunately he did not have a tape measure to check this!

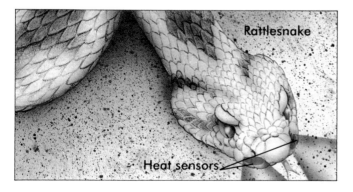

Rattlesnake

Heat sensors

Q How do rattlesnakes hunt in the dark?

A Rattlesnakes live in the Americas. They hunt rodents and other small animals. Just below their eyes, rattlesnakes have a pair of heat sensors. Using these sensors, a rattlesnake can find its prey even in total darkness.

Q Which is the thinnest killer?

A The vine snake lives in the forests of South America. It may grow to be 8 feet long, but is only about ½ inch in diameter. It preys on birds, but it will give a poisonous bite to any creature which disturbs it.

Vine snake

Q Which queen died from a snakebite?

A It is said that, in 30 BC, Queen Cleopatra of Egypt was captured by the Roman army. She knew she would be executed and decided to commit suicide instead. She asked a servant to bring her an asp - the symbol of the ancient royal family of Egypt. It was also a deadly poisonous snake, and Cleopatra knew its bite would kill her.

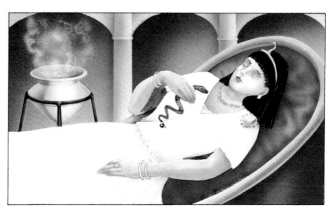

The sundew plant's sweet, sticky liquid attracts insects, and also dissolves them.

Where do poisonous plants grow?

Aethusa

Many different plants contain poisonous chemicals. Some plants use their poison to stop animals from eating them, but often the chemical is useful to the plant in other ways. Poisonous plants can be found anywhere. You shouldn't eat any plant unless it is known to be safe.

Q What is a Venus fly-trap?

A A Venus fly-trap is a plant which feeds on insects. It has special hinged leaves. When an insect lands on a leaf, it quickly folds up, trapping the insect. The plant then releases digestive juices which kill the insect and remove minerals from its body.

Foxglove

Q Can poisonous plants be used as medicine?

A Sometimes. For example, foxgloves grow all over Europe. They have large purple flowers and are often found in woodland. The leaves of the foxglove are very poisonous. Scientists can extract a chemical called digitalis from foxglove leaves, which is used to treat some types of heart disease.

Q Which common medicine comes from plants?

A In most homes, people keep aspirin tablets to cure headaches and other aches and pains. Aspirin is now made artificially, but was at first made from the leaves of a bush that grows in North America.

Venus's fly-trap

Aethusa looks like wild parsley, but is deadly if eaten, and has killed many people.

Adonis daisy

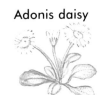

Acid from the Adonis daisy is used to treat heart disease, but is deadly if eaten raw.

Deadly nightshade

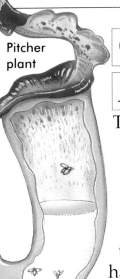

Pitcher plant

Q **How do pitcher plants kill prey?**

A Pitcher plants grow in tropical rainforests. They have cup-shaped leaves, which have a sweet liquid in the bottom to attract insects. Once an insect is inside the cup, it cannot escape because of the downward-pointing hairs in the cup. The insect drowns and is digested to provide the plant with minerals.

Q **Are all berries edible?**

A No. Berries are the soft, juicy covering of seeds. When they fall to the ground, berries provide the water and nutrients the seeds need to begin growing. Some berries are safe to eat and are very tasty. But many berries, such as deadly nightshade, are poisonous to humans. Never eat a berry unless you are sure it is safe.

Nettle

Q **Which stinging plant is good for human hair?**

A The stinging nettle is covered with stiff hairs which inflict painful stings on any creature which touches them. However, the juice of the stinging nettle is said to help human hair to grow strong and healthy.

Artist's impression of a ya-te-vau

Q **Is there a man-eating tree?**

A In the 1880s, explorers in Central and South America reported a tree-like plant called the ya-te-vau. When an animal brushed against its branches, they would wrap around the victim, pricking its skin with spikes. The blood was then absorbed by the plant. It was said that a large tree would be able to kill a human. Despite many reports, no ya-te-vau has ever been seen by scientists.

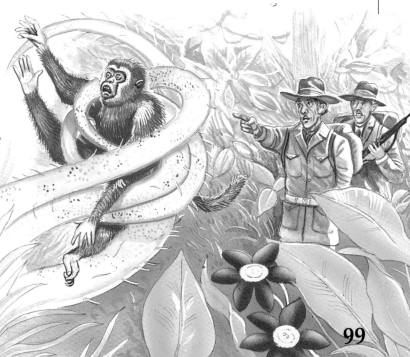

The rare and inedible Boletus purpureus turns bright blue if cut open.

Why can't you eat all wild mushrooms?

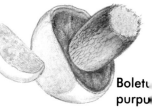

Boletu purpu

Mushrooms are types of fungus which grow in many areas of the world. Some are edible, but many are extremely poisonous. You should never eat a fungus unless you are expert at recognising the different types. Many people die from eating toadstools.

Q Which fungus has got a dangerous twin?

A There are several types of morel fungus. The esculenta morel grows in woodland in the spring and tastes delicious. The false morel grows at the same time of year and looks almost identical, but is deadly poisonous.

Q Which is the most poisonous toadstool?

A The death cap toadstool gets its name because most people who eat it die. The death cap is a white toadstool with a pale bronze cap. It produces severe stomach pains about six hours after being eaten, and quickly destroys your liver and kidney. There is no known cure.

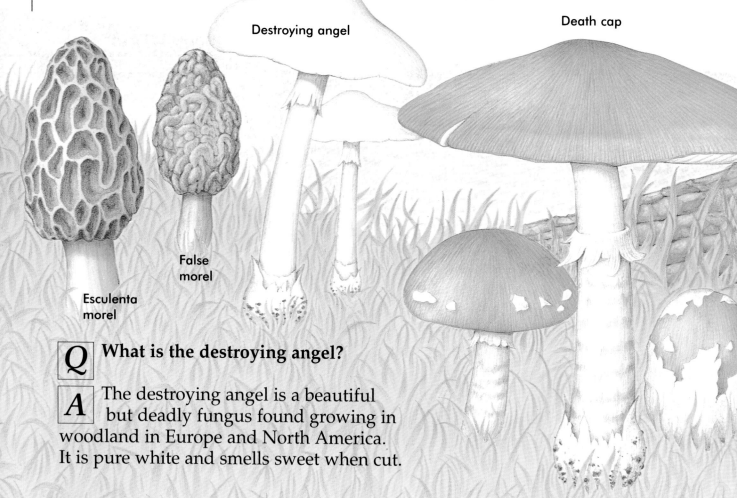

Destroying angel

Death cap

Esculenta morel

False morel

Q What is the destroying angel?

A The destroying angel is a beautiful but deadly fungus found growing in woodland in Europe and North America. It is pure white and smells sweet when cut.

100

The smelly stinkhorn smells like rotting flesh to attract flies which disperse its spores.

Stinkhorn

Artists often draw fairies with pretty red spotted toadstools. These fly agaric toadstools are very poisonous.

Q Which fungus kills trees?

A The honey fungus grows on deciduous trees, such as oaks and maples, in many parts of the world. If honey fungus spores land on a cut in a tree's bark, they will grow rapidly. Within a few months, the tree is dead, its wood rotted to white powder.

Honey fungus

Q Which is the most valuable fungus?

A The most valuable fungus is the black perigord truffle which grows underground in Europe. It has a delicious and unique flavour. Its price varies during the year, but can be as much as ten times the price of pure silver.

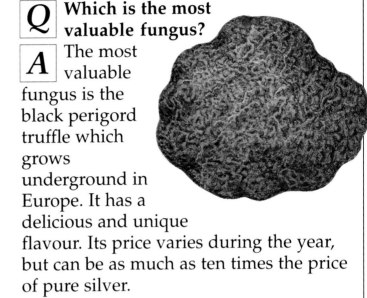

Q Which is the most edible group of fungus?

A Nearly all the types of boletus mushroom are edible. The penny bun boletus, or cep, is harvested by mushroom farmers to sell. However, the devil's boletus is poisonous.

Penny bun boletus

Devil's boletus

Which are the polar predators?

Arctic skua

The vast areas around the North and South Poles are very cold. The land and sea may appear to be bleak and empty, but many animals live there. Some of them are powerful hunters.

Q Which is the most powerful polar hunter?

A The polar bear is the most powerful and aggressive animal in the Arctic. It grows to be about 8 feet long, though some reach over 10 feet. Polar bears are very strong and hunt seals, fish, and even reindeer.

Q Which polar predator changes colour?

A During the summer, a stoat has a reddish-brown coat. This allows it to hide from its prey on the tundra (the area between the polar ice cap and where trees start to grow). In winter, its coat turns white, with a black tail-tip, so it can blend into the snowy background.

Polar bear

Q Which seal hunts penguins?

A The leopard seal gets its name because it has a spotted coat and is a fierce hunter. It likes to lurk beneath the ice of the Antarctic, waiting for penguins to dive into the water. When a penguin is within reach, the seal grabs it in its powerful jaws. Leopard seals are about 10 feet long.

Leopard seal

The stoat kills by leaping on its prey.

Stoat

Polar bears smash through the ice to reach sleeping seals.

Musk oxen

Q **How do musk oxen protect themselves?**

A Musk oxen are shaggy sheep-like animals which live on the Arctic tundra. When a pack of wolves approaches, the musk oxen bunch together in a circle with their horns facing outward, and their young in the middle. The wolves cannot break through this barrier, and usually go away.

Q **Which hunter travels a long way in search of prey?**

A The albatross hunts fish and crustaceans (sea-creatures with hard shells) in the seas around the South Pole. It travels over the seas for months at a time, only coming to land to lay eggs. An albatross may fly over 50,000 miles each year.

Albatross

Walrus

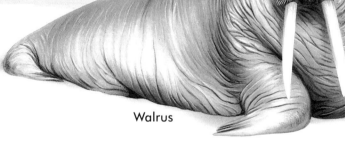

Emperor penguin

Q **Which predator eats only shellfish?**

A The Arctic Ocean is home to vast numbers of shellfish. Walruses, which grow to over 10 feet long, eat shellfish. Walruses live in large herds, but hunt on their own. They have long tusks. Nobody is quite certain how a walrus eats its prey, but it seems to suck the fish out of their shells.

Q **What does the largest penguin hunt?**

A The largest penguin is the emperor penguin, which stands over 3 feet tall. It has a long, slender bill which is good for catching slippery prey. Most of the time it hunts squid, and occasionally takes fish or other sea animals.

The flower mantis looks exactly like a flower, but it kills insects which come within range.

Flower mantis

Many hunting animals disguise themselves. This allows them to get closer to their prey and so stand a better chance of making a kill.

Q Which hunter uses a 'worm' as bait?

A The alligator snapping turtle lies on the beds of rivers and lakes in North America with its mouth gaping wide open. When a fish swims past, the turtle wiggles its tongue, which looks just like a struggling worm. If the fish approaches, the turtle snaps its mouth shut on the unfortunate victim.

Alligator snapping turtle

Q Which snake is almost invisible?

A The eastern green mamba lurks in trees in southern Africa. Its bright green colour is identical to that of the leaves on the trees where it lives. This makes the mamba almost impossible to see, unless it moves.

Q Which larvae catch ants?

A Ant-lion larvae feed on ants and other small insects by setting a clever trap. The larva digs a tunnel in sand and then buries itself at the bottom. To a passing ant, this appears to be just a small hole. But if the ant climbs, or falls, into the hole, it is pounced on by the waiting ant-lion larva.

Ant-lion larva

Q Which spider opens a door?

A Trapdoor spiders take their name from the way they catch their food. They dig a tunnel and make a door at the surface end. The door is made of silk mixed with sand and soil. The spider waits just beneath this door. When its prey walks past, the spider leaps out and grabs its victim.

Trapdoor spider

Hover flies are coloured yellow and black to look like wasps, but they cannot sting.

Hover fly

The Champawat man-eating tiger killed 436 people in India.

Eastern green mamba

Q Which great cat has a coat which allows it to hide in the shadows?

A The tiger hunts its prey through dense jungles and grasslands. Its striped black and orange coat helps it to merge into the bright sunlight and dark shadows of its home. A tiger stalks slowly through the undergrowth until it is close to its prey. Then it springs and kills the prey with a bite to the throat.

Q Which fish goes fishing?

A The angler fish lies on the bottom of the sea, partially buried in sand. It has a spine over its mouth which it uses like a fishing rod. On the end of the spine is a flap of skin, which looks a crustacean or worm. This acts as bait, and is waved about to attract small fish. When its prey is near, the angler opens its huge mouth, sucking its meal inside.

Q Which fish electrocutes its victims?

A The electric ray buries itself in mud at the bottom of the sea. When a small fish comes near, the ray emits a powerful electric charge to stun the fish, allowing it to be eaten with ease.

Electric ray

Angler fish

Gannets dive into water from a height of 30 metres to catch fish.

Gannet

Many different types of bird and bat fly in search of prey. Some, such as swallows and bee-eaters, which eat insects, catch their prey in flight. Others swoop from the sky to catch prey on the ground. Both techniques require superb eyesight and great flying skill.

Q Which is the heaviest flying hunter?

A Condors, which fly in the Andes mountains of South America, weigh up to about 26 pounds and have a wingspan of 9 feet. Condors feed mainly on carrion (dead animals), but are also thought to take small mammals.

Andean condor

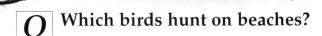

Monkey-eating eagle

Q Which eagle hunts in forests?

A The monkey-eating eagle of the Philippines is one of the largest eagles. It is about 32 inches long, with short, powerful wings. These are small enough to allow it to dart through branches in search of the monkeys and other mammals on which it feeds.

Q Which birds hunt on beaches?

A The great skua of the North Atlantic Ocean finds much of its food on beaches and close to the coast. It also hunts further inland. Skuas hunt the eggs and chicks of other sea birds, and attack any crabs or shellfish which they can find. Many gulls hunt on beaches too.

Great skua

Q Which birds hunt other birds?

A The most successful bird-killers among the birds of prey are falcons. They catch small birds by chasing them at high speed. Falcons can twist and turn while flying very fast. They dive on larger prey from above.

108

The black skimmer
flies with its lower
beak in the water,
hoping to strike a fish.

Black
skimmer

King vulture

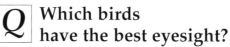

Q How do bats find insect prey?

A Most bats fly about to hunt insects. They find their prey in the darkness of night by using a form of sonar, called echo-location. The bat makes high-pitched clicking sounds which bounce off objects. It can tell the size and position of an animal by its echo, and can then home in for the kill.

Q Which birds have the best eyesight?

A Eagles and hawks have good eyesight to enable them to catch their prey, but the sharpest eyesight probably belongs to vultures. These birds soar about 6,000 feet above the ground, looking for food. Even from this height, the birds can see a decaying carcass and swoop down to feed.

Vampire bat

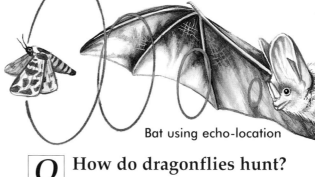

Bat using echo-location

Q Which bat sucks blood?

A Vampire bats are about 4 inches long, and live only in South America. They hunt at night, landing close to their prey and then crawling forward. They bite their victims and then use their tube-shaped tongue to suck up the blood.

Q How do dragonflies hunt?

A Dragonflies prey on other insects, usually gnats and flies. They have very good eyesight and can spot a flying insect 125 feet away. The dragonfly swoops on its prey, using its legs to catch it, and kills it with a single bite.

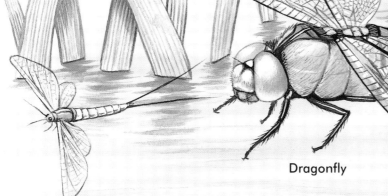

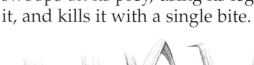

Dragonfly

109

Pronghorn antelopes can run as fast as 35 miles an hour to escape hunters.

Why do lions rule the plains?

Pronghorn antelope

The lion is the largest and most powerful hunter on the African plains. It may grow to reach 10 feet in length and weigh 660 pounds. Few animals are able to survive an attack by a lion.

Lion

Lioness

Elephant

Q How do elephants protect themselves?

A On the African plains, elephants live in herds of up to fifty animals. Although fully-grown elephants are too large to be hunted by lions and leopards, young calves are in danger of being attacked. If a predator is seen, the older females will charge forward, trumpeting loudly, to drive it away.

Q How do lions hunt?

A Lions prey on zebras, gnu and various antelopes. They also eat birds and other small animals if they are very hungry. Lionesses do most of the hunting, working together to drive prey toward a place where the other lions are hiding.

Q What is a pride of lions?

A Lions live in family groups called prides. Each pride is made up of an adult male, up to fifteen females (lionesses) and their young, and sometimes young males too. The pride lives and hunts together.

110

Hyenas have jaws strong enough to snap and devour a buffalo thigh bone.

Lions make a kill on only one-third of their hunts.

Q Which animal steals meat from lions?

A After lions have made a kill, they feed immediately. Sometimes, hyenas try and steal their food. A large pack of hyenas can drive a small number of lions away from a carcass.

Kangaroos

Hyena

Rhinoceros

Q How do kangaroos fight?

A In Australia, kangaroos live on the open plains and are able to flee very quickly when danger threatens. However, when they cannot escape by running, kangaroos will lash out with their powerful hind legs. A single kick can knock a dog unconscious.

Q Why are rhinoceroses dangerous to humans?

A Rhinoceroses eat plants and do not hunt other animals, but they can still be very dangerous. When a rhinoceros feels threatened, it will lower its head and charge. The long horn on its forehead is a very dangerous weapon, especially because two tons of angry rhino lie behind it!

Which killers lurk in the woods?

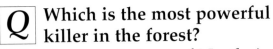

Forests can be dark, forbidding places. Undergrowth may conceal a large animal and strange sounds echo through the trees. Hunters who lurk in the forests are among the most powerful in the world.

Q Which is the most powerful killer in the forest?

A The grizzly bear of North America is a huge and powerful hunter. It is usually about 8 feet long, but some grow to over 10 feet. They hunt fish and kill large deer and bison if they get the chance. Grizzlies usually avoid humans, but have been known to kill them.

Q Are any forest mammals poisonous?

A The short-tailed shrew of North America can deliver a venomous nip. It hunts vast numbers of worms, insects and small mammals, which are injected with fatal poison from its lower incisor teeth.

Short-tailed shrew

Q What is a bobcat?

A The swamps and forests of North America are home to the bobcat. Bobcats are about three feet long and have a beautiful spotted coat. They used to be hunted for their fur. Bobcats prey on rabbits, rats, and other mammals.

Bobcat

Grizzly bear

Wolverine

Q Which killer is called "the glutton"?

A The wolverine of North America is known as the glutton (meaning greedy) because of its habit of killing more than it needs, and storing the food in snow. An adult wolverine can kill a reindeer and will return to the carcass many times until it has all been eaten.

Leopards carry their prey into trees to eat undisturbed.

Bobcats often hook fish from streams with their front paws.

Red fox

Q Which forest hunter lives in cities?

A The red fox is found across Europe, Asia, and North America. It usually lives in forest areas, where it hunts rabbits, birds, and insects. However, in recent years, the red fox has begun to live in cities, where it preys on garden birds and raids garbage cans for food.

Q Which forest hunter kills silently?

A Leopards are famous for their ability to glide silently through the night. They use this skill to creep close to prey and seize them in their jaws. In 1922, a leopard killed a man so quietly that a second man, sitting just two metres away, heard nothing at all.

Q Which is the rarest forest killer?

A This may be the Tasmanian wolf of Australia, if any still exist. The Tasmanian wolf is about the size of a big dog, up to about four feet long. It preys on kangaroos and other medium-sized creatures.

Tasmanian wolf

Q What is the Tasmanian devil?

A The forests of Tasmania in Australia are home to a creature which earned itself the name Tasmanian devil because it fights savagely when cornered. This stocky, muscular creature is about three feet long and preys on birds, lizards, and other small animals.

Leopard

Crocodiles often store meat under sunken logs, returning when the flesh has begun to rot.

Many river animals live by attacking and eating other creatures. Some eat almost everything in sight, preying on anything they can reach. Only a few are able to kill animals as large as a human.

Q Which crocodile is most dangerous to humans?

A The salt-water crocodile of Australia and southeast Asia can grow to reach 20 feet in length. These crocodiles frequently attack any humans they come across. During the Second World War, about 900 Japanese troops were trapped in a swamp. After an attack by several crocodiles, only 20 of the men were left alive.

Salt-water crocodile

Q Which river fish can reduce an animal to bones in minutes?

A Piranhas are small fish with razor-sharp teeth, found in the rivers of South America. They live in large groups. Piranhas are very aggressive, and if they scent blood, they will even attack large animals. Humans can be reduced to skeletons in minutes.

Q What is the 'death roll' of a crocodile?

A Crocodiles have simple jaws which can only open and shut, not chew. When a crocodile grabs a creature too large to swallow whole, it will drag its victim to the bottom of the river and spin very quickly. This 'death roll' drowns the prey.

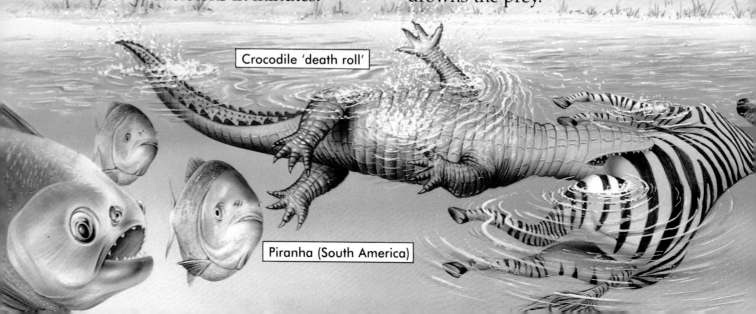

Crocodile 'death roll'

Piranha (South America)

The Ganges shark can attack people who bathe in the River Ganges in India.

Large pike drown ducks by gripping their feet and pulling them under the water.

Q Which insect eats fish?

A The most voracious hunter in European ponds is the nymph (larva) of the dragonfly. The nymph has long, clawed mouth-parts which it shoots out to spear prey. When it first hatches, a nymph preys on microscopic creatures, but eventually it will spear tadpoles and small fish as food.

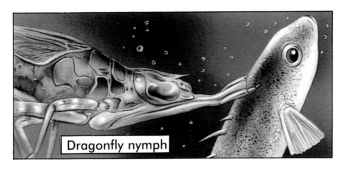

Dragonfly nymph

Q How do crocodiles ambush prey?

A When deer or other animals approach a river to drink, they may be in danger. If a crocodile sees them, it swims underwater to within a few feet of its victim, then rushes forward to grab the animal in its jaws.

Q Which fish "shoots" its prey?

A The archer fish lives in the mangrove swamps of southeast Asia. It was given its name because it shoots a stream of water at insects resting on leaves near the water. The archer fish can hit insects up to 12 feet away, sending them tumbling into the water to be eaten.

Crocodile

Archer fish (South-East Asia)

115

Tyrannosaurus
had teeth six
inches long.

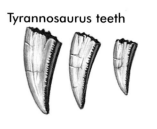

Tyrannosaurus teeth

Can fossils reveal killers?

Scientists who study the fossilised bones of ancient creatures are called palaeontologists. By studying these preserved bones, palaeontologists can work out how an animal lived and what it looked like. Different kinds of bones are found each year, so our knowledge is increasing steadily.

Q Which was the largest land hunter ever?

A The largest and most powerful predator ever to walk the Earth was the dinosaur *Tyrannosaurus rex*, which lived about 80 million years ago in North America. This creature was about 50 feet long and walked on its hind legs. Some people think that it ate carrion (dead animals) rather than hunting live prey.

Q Which dinosaur used a claw to kill?

A *Velociraptor* was a dinosaur which lived in North America about 75 million years ago. It was able to run very quickly on its hind legs. It killed other dinosaurs by gripping them with the claws on its front legs, and slashing with the large claw on each of its hind legs.

Velociraptor

Dimetrodon

Q Which animal had a sail on its back?

A *Dimetrodon* was an early reptile which lived about 280 million years ago in North America. It had a tall fin, like a sail, on its back. The fin was used to absorb heat from the sun very efficiently. It used its long teeth to attack other reptiles.

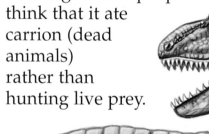

Tyrannosaurus

Ichthyostega

The first land predator with a backbone may have been Ichthyostega, an amphibian which lived 400 million years ago in Greenland.

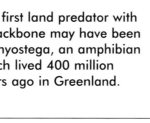

Q Which giant bird hunted mammals?

A About 15 million years ago, *Phorusrhacus* stalked the grasslands of South America. It was about ten feet tall, and preyed on mammals and reptiles which it chased on its long, powerful legs. It had a large, hooked beak.

Phorusrhacus

Q What was the "Jurassic tiger"?

A *Allosaurus*, known as the "Jurassic tiger", was a powerful hunting dinosaur which lived about 140 million years ago in Australia, Africa, and North America. It was 35 feet long and preyed on other dinosaurs.

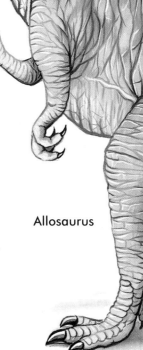

Allosaurus

Q Why did the sabre-tooth tiger have such large teeth?

Smilodon

A *Smilodon*, often called the sabre-tooth tiger, lived about 50,000 years ago in America. It used its extremely long teeth to stab the large mammoths and bison on which it preyed.

Q Which dinosaur hunted fish?

A In 1983, the remains of a flesh-eating dinosaur, called *Baryonyx*, was found in England. This dinosaur was about 30 feet long and had a large claw on its front leg. Scientists think that the *Baryonyx* hooked fish with this claw and then secured them in its many small teeth.

Baryonyx

Hypsilophodon

INDEX

Created by Zigzag Publishing, a division of Quadrillion Publishing Ltd, Godalming Business Centre, Godalming, Surrey GU7 1XW, England.

This edition contains material previously published under the titles Zigzag Factfinders/100 Questions & Answers - *Fantastic Sea Creatures, Minibeasts, Monster Animals,* and *Dangerous and Deadly.*

Color separations: RCS, Leeds, England, and ScanTrans, Singapore
Printed by: Tien Wah Press, Singapore

Distributed in the U.S. by SMITHMARK PUBLISHERS a division of U.S. Media Holdings, Inc., 16 East 32nd Street, New York, NY 10016

Copyright © 1997 Zigzag Publishing
First published in 1993 by Zigzag Publishing Ltd

All rights reserved. No part of this publication may be reproduced, stored in a retrieval system or transmitted by any means, electronic, mechanical, photocopying or otherwise, without the prior permission of the publisher.

ISBN 0-7651-9338-8
8459